evaluations. The sales department's mission is to develop sales plans to take advantage of the existing marketing programs and to sell profitably as much of the product as can be produced.

Within the research department, specific areas, for example, an analytical group, will have even more specific definitions of mission, which support the overall mission of the department. The lower in the organization, the more narrow the definition of mission.

While most of the above may sound extremely elementary, failure to have a clearly defined mission results in poor performance of managers and the failure of organizations. Furthermore, without a clear correlation of the missions of the various subunits, the organization's objectives are unlikely to be achieved.

Once the mission is clarified the planning stage begins. Planning is a cyclical exercise that is composed of discrete steps. The steps in a typical cycle are shown in figure 1-1.

LONG-RANGE PLANNING

The definition of long-range planning varies not only among organizations but may also vary widely within an organization. Generally speaking, within a research organization, long-range planning encompasses that time period required to conceive, complete development

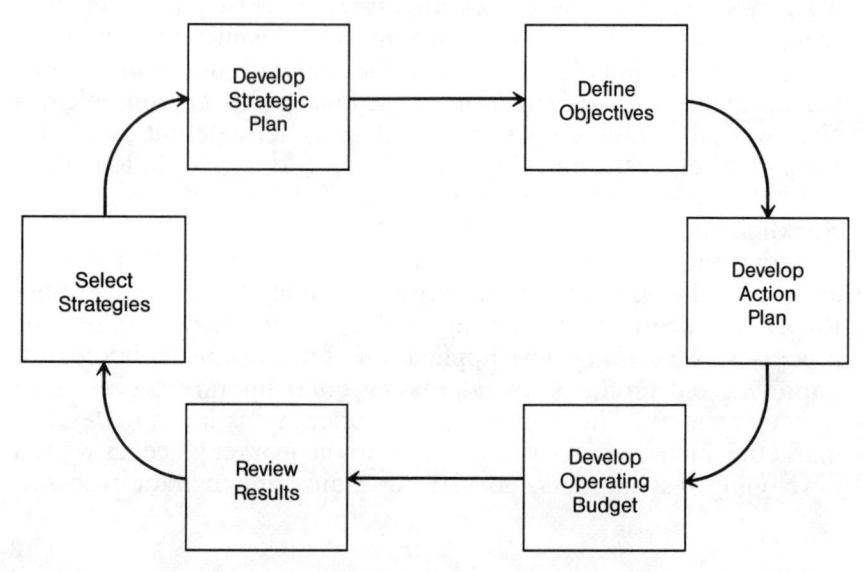

Figure 1-1. Planning cycle.

of, and introduce a new product to the market, or to solve a specific problem. This period can vary widely from organization to organization. The pharmaceutical and agricultural chemical industries, as well as those in the area of basic research, tend to have a long-range planning period of from seven to twelve years. Long-range planning for the finance department of an organization is generally in the four-to-five-year range. This presents the R & D manager with a dilemma, since the products or problems on which he is working may not be expected to have a positive impact on the organization's performance during the period for which the financial plan has been developed. This also challenges the R & D manager to set up extremely good communications with other departments in the organization so differences in planning periods can be accommodated. Some significant problems in linking the long-range plans of organizations with those of research departments are pointed out by Weil and Cangemi (1983). These problems include a lack of clearly defined corporate objectives; a lack of effective ways to identify corporate needs; poor communications; and the corporate management's adversity to high risk R & D projects.

STRATEGIC PLANNING

Strategic planning focuses on the broad policy questions facing an organization. These include: Where are we now? Where do we want to be? How do we get there?

A formalized strategic-planning process consists of the following steps:

1. Definition of basic mission or purpose
2. Evaluation of current position in a business life cycle or in the competitive scientific community
3. Evaluation of relative strengths and weaknesses
4. Evaluation of relative competitiveness
5. Review of methods for adding new products or services (internal development or external acquisitions)

Additionally, it is necessary to make assessments and forecasts relating to external factors, such as the general economic situation, the effects of regulation, and inflationary or deflationary trends.

A survey of the current status of the various products or business segments is required to determine where your organization stands in relation to competitors. Relative strengths and weaknesses may be assessed through several methods, most of which require generation of a matrix to determine strategies to be employed (fig. 1-2).

outside contractual service should be considered to conduct specific aspects of the work for you. The decision to go outside may have a significant impact on the budget for a specific project.

Capital Resources

Capital equipment required to complete the projects assigned to your laboratory should be included in a separate capital expenditures plan, as discussed in chapter 1. In developing your action plan, you should also take into account the capital equipment needed and the time period in which this equipment will be required. It does a researcher little good to discover that the gas chromatograph that he needs to complete his project in March will not be available for purchase until late in the year because of budgetary considerations.

Major pieces of capital equipment generally have long delivery dates. Lead times of four to six months are not uncommon for delivery of items such as NMRs (Nuclear Magnetic Resonance Spectrometers), mass spectrometers, and even laboratory furniture. Timeliness in obtaining needed capital equipment will directly affect the project schedule.

Support Resources

These resources fall into two categories: support required as a part of the project charged directly to the project, and general support provided by the organization.

Direct Project Support. The amount of support services required for each project will vary considerably depending on the nature of the work being conducted. For example, a chemist assigned to developing new formulations of existing products may require the equivalent of one to one-and-a-half analytical chemists to analyze the work he is generating. A process chemist who does a great deal of his own analytical work in following the course of the reaction may require substantially less analytical support.

It is your responsibility to determine what support services your personnel require and to make the necessary arrangements to assure that this support is available when required. Since the personnel providing this type of support probably will not be under your direct supervision, you will have to maintain close communication with the supervisors of the areas providing that support so that your needs are taken into consideration in their planning.

Indirect Support. Support that may not be charged directly to a project can be provided through the R & D department or may be provided by other departments within the organization. Examples of this type of support include

- General laboratory supplies (such as gases, glassware, and apparatus)
- Laundry service for laboratory coats
- Waste disposal
- Technical information services
- Purchasing
- Patents
- Personnel
- Computer services
- Accounting

Many organizations have assigned administrators to be responsible for providing the services mentioned above as well as other services. However, it is still your responsibility as a manager to make sure that the support required is both readily available and is in fact being received by your personnel.

Facility or Space Resources

Additional space requirements or special facilities, such as animal-handling laboratories, should be identified in your action plan. These requirements are the most difficult to provide within an existing laboratory and will require the longest lead times to obtain, especially if new construction is required.

ORGANIZATIONAL STRUCTURE

The structure to be used to satisfy your objectives should be specified in your operating plan. There are several types of organizational structures that can be set up to achieve the results required from your laboratory.

Project Teams

Within the laboratory, a project team generally is composed of a group of researchers with different backgrounds assigned to complete a multifaceted project. This group could include synthetic organic

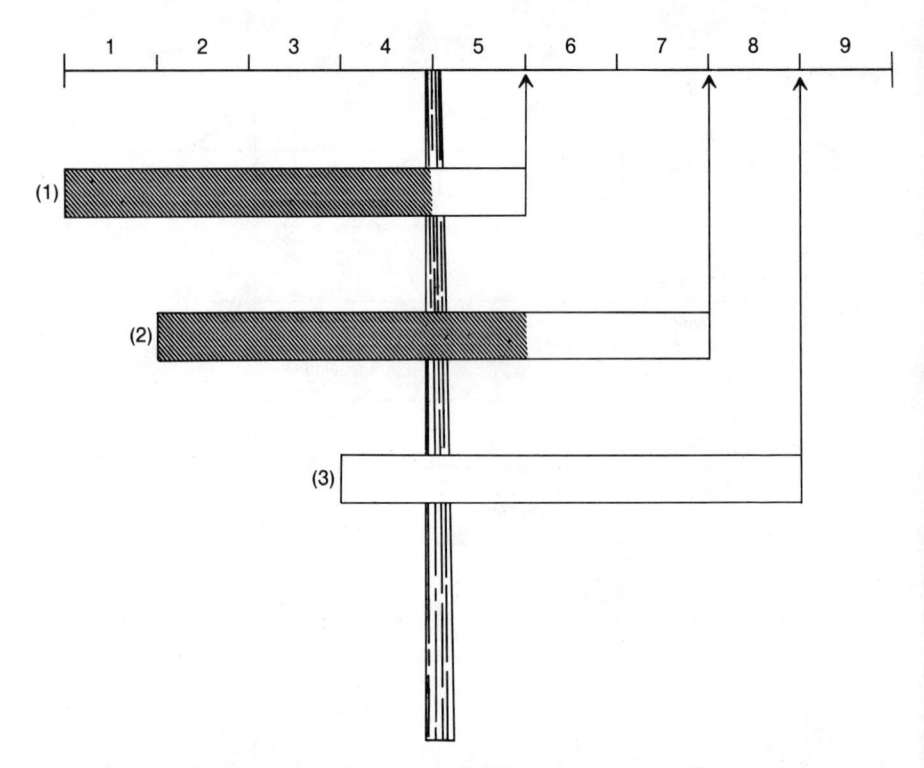

Figure 3-2. Milestone analysis (week). The bars represent laboratory activities: (1) cell culture studies; (2) biochemical analysis; and (3) animal studies. When reviewed at the end of week 4, (1) is on schedule and due for completion at the end of week 5; (2) is one week ahead of schedule and should be completed at the end of week 6 rather than week 7 as initially projected; and (3) was not started as planned and is at least one week behind the scheduled completion time.

at the end of period 7. Milestone analysis indicates that these studies are ahead of schedule and will be completed at the end of time period 6. The animal studies were to have begun at the beginning of time period 4. However, at the beginning of period 5 it is clear that they were not begun on schedule and are a week behind. Milestone analysis allows the manager to get an overview of project progress without the detail provided with the Gantt chart. This type of analysis might be more suited for control at a higher management level than in the laboratory, because at the latter level information on technical aspects of an individual project's progress is more vital.

Network Analysis

The preceding simple, but effective, methods of time-event analysis have provided the basis for the more sophisticated control techniques of network analysis—among which CPM (Critical Path Method), PERT (Program Evaluation and Review Technique), IMPACT (Implementation Planning and Control Technique), PRISM (Program Reliability Information System for Management), and GERT (Graphical Evaluation and Review Techniques) are but some variations. Perhaps the most widely used method is PERT, which was developed for the U.S. Navy by the management consulting staff of Booz, Allen and Hamilton to control the research, development, and production phases of the Polaris Weapon System project in 1958.

PERT. The first step in this network analysis system is to establish all major milestones, or events, that must be performed in order to complete the program. The milestones are then sequenced according to their technological and/or administrative interdependence. Events, represented as circles, are connected by arrows indicating the direction of the activity between sequential events (fig. 3-3). Each activity, of course, has a beginning and end (milestone) and may be performed in series or parallel depending on the interdependence of the individual events.

Working through the logical progression of events in the network, three time estimates are assigned initially for the completion of each activity. These estimates include (1) the most optimistic time required to complete each program event, T_O; (2) the most likely time for completion, T_M; and (3) the most pessimistic time, T_P. The expected time, T_E, is then computed for each activity using the following formula:

$$T_E = 1/6\,(T_O + 4T_M + T_P)$$

With linear programming methods, the network can then be analyzed for the most time-consuming path—the *critical path*. Timely completion of the events on the critical path is requisite for completion of the program on schedule and thus affects control over the program. Once the critical path problem has been solved, the manager is able to see immediately which event is throwing the project out of control. This information can then be used to reallocate resources or implement other remedial steps to reduce the critical path. One potential remedy for critical path activities that are running behind schedule is to reallocate resources from *slack paths*.

Normally, the earliest project completion time is expressed as the sum of the T values of each activity along the critical path. This

The key is an active committee. Two guidelines for creating an active safety committee will help the laboratory manager. First, select a few people who are self-starters and understand safety to help give the committee a rolling start. Second, give the committee some freedom to act. The committee can write proposals, procedures, policies, set up training courses, and arrange for films. However, if the committee is perceived as not having any responsibility, the laboratory personnel will think of it as a showpiece and not as being functional. Under these circumstances, even the most ardent and active members will soon tire and lose interest. A common factor among successful safety committees is leadership and commitment by line management.

Accident Investigation and Accident Prevention

Articles on the importance of accident investigation conclude that the most important reason for investigating accidents, and the chief benefit to be derived from the investigation, is accident prevention (Steere 1971) and avoidance of the resulting injury, illness, or property damage. A good choice to investigate an accident is an individual who supervises workers in the laboratory. That individual is familiar with the laboratory procedures and has first line supervisory responsibility, as well as a specific responsibility to prevent accidents. Whether the investigator is a supervisor, the laboratory manager, or a team assigned from the laboratory safety committee does not matter so long as the purpose of accident investigation and the advantages of a successful investigation are clearly understood. The investigator should have frequent contact with the equipment and procedures and thus be able to distinguish between an equipment problem and an improper action. The accident investigation report is submitted to the safety committee to study ways to improve performance, to the supervisor for personnel discussions, and to the medical department for occupational health statistics. The authority to act rests with management, not with the assigned investigator.

The accident investigation focuses on conditions that could endanger lives, cause damage to equipment or facilities, or destroy experimental results. Being in the immediate area, a supervisor can take corrective action swiftly. The results of the accident investigation can be communicated quickly to other individuals, to alert them to any potential problems. There are also benefits for supervisors. Prompt and thorough investigation is evidence of concern to employees and helps in winning their trust and respect. Effective accident investigation can increase productivity and reduce costs by minimizing lost time due to accidents. Investigation and the resulting accident prevention become tangible evidence of ability, efficiency, and managerial capability; and this can be important for the career path of the individual.

There is no such thing as an unimportant accident. Though the immediate results might be classified as minor, an accident itself cannot be classified as unimportant. If an accident goes unreported, it cannot be investigated, and its causes cannot be corrected. It then becomes a time bomb waiting to be triggered again, and the odds are that the next time the results could be far more serious or even fatal. Therefore, every employee must be urged to report all accidents or near accidents for investigation. Many reasons are given for not reporting an accident:

"It was near quitting time and I did not want to miss my ride."
"I did not want to spoil the safety record."
"I did not want to go all the way to the medical department for treatment."
"I was afraid my fellow workers would kid me."
"My supervisor gets mad if I ask for a pass to go to the clinic."
"I did not want to lose any time from my job."
"I did not want the investigating committee asking a lot of questions."

There are several things the supervisor can do to ensure that minor accidents are reported. First, use the orientation program to help employees understand why minor accidents and near misses should be reported. Train the employees to go to a supervisor to discuss that near miss. In many near-miss incidents, the only reason it did not become an accident was because the employee's reflexes were good, or no one was standing nearby, or the person was just lucky. Help the employee understand that situations cannot be corrected unless they are reported, and that eventually an accident might happen that could have serious results.

If an employee does report an accident or a near miss, treat the report as important. Tell the employee that you appreciate his responsiveness, and treat each report with respect and gratitude. If employees believe that accident reporting is important to management, they will generally respond favorably. Do not chew them out or give them bad reports because of the accident, but treat it as a learning experience and discuss ways of correcting the situation. The supervisor should start immediately to investigate and take the corrective actions appropriate to their professional activities. Ensuring a good information exchange might mean bringing together parties who have experienced some failure in communications. Whatever the problem is, a supervisor must be aware that it is not likely to go away and should be addressed as soon as it is recognized. If necessary, the research manager should work to further develop his own skills, for his responsibilities will only increase and his efforts will enhance his potential for advancement.

This chapter is designed as a general guide for the manager in helping personnel in the department become better communicators and in enhancing and expanding his own skills. The practical hints included here are not intended to take the place of a thorough study of the skills that go into effective writing or speaking. The bibliography at the end of the chapter suggests a number of references, instructional guides, and books for self-improvement that may be useful in this regard. Both supervisors and employees should remember that it takes practice to keep communication skills current.

SEMINARS AND PUBLICATIONS AS AN INCENTIVE FOR EMPLOYEES

Every laboratory manager searches for methods of improving working conditions for his people. Often the research manager does not have direct control over such tangible and intangible factors as the working environment, compensation levels, or the types of benefits offered. He can, however, provide incentives for his employees in other ways. One of these is by encouraging his employees to participate in professional meetings, submit papers, or give seminars.

The employee can write papers for professional journals in his field of expertise or hold seminars to gain peer recognition. He can write articles for lay publications or give talks to lay organizations to promote more widespread understanding of technical subjects. These activities will improve the employee's morale and add purpose to his profession and his job. Getting employees involved in additional activities enhances their self-esteem and generally improves overall job performance. Practicing the communications skills required for seminars and papers also helps the employee to use those same skills at his job and provides a greater potential for career advancement. The laboratory manager will in turn see a general improvement in daily communications skills and the quality of project reports as the employee has an opportunity to practice the skills he has learned.

WRITING RESEARCH REPORTS

An important objective for an employee writing a research report is to provide enough information so that the supervisor, research manager, head of the business, government regulator, or civil authority can be informed about the topic and thus make a decision when one is required. For this to take place, reports must be accurate, clear, as brief as possible, and contain sufficient technical information so that a judgment can be rendered in a timely manner (Furnas 1948).

We can divide most reports into three broad categories: administrative (including financial) reports, technical records, and publications. Administrative reports are those that provide a level of information sufficiently detailed for a decision to be made. Technical records include laboratory notebooks or other documents that provide a permanent record of experimental activities and results. The last type of reports are those prepared for publication, which are covered in another section of this chapter.

It is important that technical activities be recorded. Since the laboratory manager is accountable for the activities in his unit, the quality of those reports directly reflects his management capabilities. An employee writing a report should be made aware that this aspect of the job offers a means of judging the quality of his work. It is also important, from a legal standpoint, to have complete documentation of activities that have been carried out in a laboratory as they occur. This is explored in more detail in chapter 6.

Rapid and accurate reporting of results helps the overall efficiency of any organization. If a decision needs to be made in a timely fashion, any delays can be very costly. If information can be shared with another laboratory to help provide a solution to an existing problem, every day of delay can add to costs through inefficiency and wasted laboratory time. The laboratory manager should be sure that his employees develop the proper skills for report writing, that they report activities promptly, that the necessary administrative and clerical services are provided, and that the reports are disseminated without delay to those who need the information.

PUBLISHING RESEARCH RESULTS IN JOURNALS

Publication of research results in professional journals is a long-established and effective method of gaining peer recognition for scientific achievement. The supervisor should make every effort to significant invention is rare, but a subjective intent to abandon can be inferred from an unreasonable delay in developing an invention after it has been made. In interference practice, which will be discussed in some detail later in this chapter, it has been held that a delay of more than eighteen months between the reduction to practice of an invention and the filing of a patent application describing it can create a rebuttable presumption that the invention was abandoned. In order to rebut the presumption of abandonment, the inventor must, if he can, come forward with legitimate reasons to justify excusing the delay. The lesson for the laboratory manager is, of course, to proceed expeditiously in the pursuit of patent protection once an invention is reduced to practice.

The third member of the trilogy of patentability requirements is that the invention be unobvious. Section 103 of the patent statute reads as follows in its relevant part:

> A patent may not be obtained . . . if the differences between the subject matter sought to be patented and the prior art are such that the subject matter as a whole would have been obvious at the time the invention was made to a person of ordinary skill in the art . . . (35 USC 103)

What makes this provision so difficult to deal with is the inherent subjectivity of the unobviousness standard. What appears obvious to one "person of ordinary skill in the art" may appear unobvious to another. In applying this standard one must take great care to avoid using hindsight, because what was unobvious "at the time the invention was made" may appear quite obvious after the problem has been solved. The obviousness issue is addressed by comparing the invention with all of the relevant prior art; and it makes no difference whether or not the inventor was aware of this art at the time the invention was made. Some inventions so clearly depart from what is known that the obviousness issue will not come up. However, many commercially significant inventions involve improvements upon known technology that can appear obvious with hindsight to one wishing to defeat the invention's patentability. Naturally the closer an invention is to the prior art, the more obvious it looks. At some point it becomes necessary to lend objectivity to this subjective test for patentability we characterize as unobviousness. In the chemical arts, compounds that differ only by a single CH_2 group are referred to as next adjacent homologs; and the existence in the prior art of an adjacent homolog will render a novel compound prima facie obvious. At one point in the development of our patent law such a showing of prima facie obviousness defeated patentability of the adjacent homolog and that was that. The enactment by Congress of the previously referred to Section 103 in 1952 resulted in a change of thinking in this area. Keeping in mind that the statute refers to obviousness of the "subject matter as a whole" the courts instructed the Patent and Trademark Office (PTO) to allow applicants to present evidence of unexpected superior properties of the chemical compound for which patent protection was being sought. Thus in a 1963 case, an applicant who was able to show that his novel triethyl substituted compound possessed anti-inflammatory properties, which the prior art trimethyl compound did not, was able to obtain patent protection for the triethyl compound. Later cases have held that even where the prior art compound possesses the property being compared, the new compound can be patented provided it is shown to be clearly superior to the prior art compound. This line of reasoning is not necessarily limited to the chemical arts; but due to the unpredictability of chemistry, and in particular when biological functions are involved, the showing of unexpected properties is of greater use to the chemical patent practitioner than those involved in the mechanical or electrical arts.

An applicant relying on unexpected properties to establish patentability of his invention will usually bring the results of comparative tests of his and the prior art compound to the attention of the PTO by way of a sworn affidavit. The affidavit will typically be prepared by a patent attorney, who can draw on prior experience to avoid the pitfalls inherent in the use of affidavits for patent procurement. Great care must be taken in preparing such an affidavit so that it does not mislead the patent examiner either through what it says or fails to say. Misleading by omission is more common than outright fabrication. A 1970 case involved a patent covering the selective herbicide 3,4-DCPA, which had been compared via affidavit with the prior art adjacent homolog 3,4-DCAA. The patent application was allowed based on a showing of the clear superiority of 3,4-DCPA as a selective herbicide. However, when the patentee sued an infringer, the court proceedings resulted in the production of the raw data that were used to prepare the affidavit. With all the data in, it was not so clear that 3,4-DCPA was superior because some tests tending to show that 3,4-DCAA had herbicidal activity with respect to certain weeds had been omitted. The patent examiner, having no investigative authority or capability, had no way of knowing of the omission and allowed the application. The court hearing the infringement suit found that the patent was invalid and unenforceable because of the patentee's intentional withholding of material facts during the prosecution of the patent applicative agencies. The resubmission of these materials to the agency marks the initial step in the *budget preparation and review* process. At this stage there exist three major intervention points where the agency may influence the final outcome of its budget. After review of the OMB guidelines, which contain ceilings for its various programs, the agency may choose to submit to OMB estimates including "over ceiling" options. This precipitates another round of OMB review, with consultations with various elements of the executive branch. At the same time, adjustments to the FY 198X budget may be made to reflect final congressional action on the FY 198(X−1) budget. Once again, agencies and OMB prepare detailed budget materials, with respect to the latest changes, which will become a part of the president's published budget document as well as the agency's justification to Congress. Even as the president submits to Congress his current services budget for FY 198X, as well as that for the four subsequent years (the five-year budget plan), OMB is conducting an internal "director's review" of the agency's budget. The OMB director then makes recommendations to the president, who makes specific budget decisions. After these are communicated to the agency, through OMB, the agency may wish to protest these decisions through "reclamas." These are usually settled by the preparation of joint issue papers by OMB and the agency. Sometimes reclama issues are settled in discussion with the OMB director and the president, but final budget decisions always rest with the president. After these decisions are made, the agency distributes dollar allowances to specific programs, making any necessary changes, and incorporates this material into the president's budget document and the agency's congressional justifications. The president's FY 198X budget and the agency's FY 198X authorization bills are submitted to Congress in January of 198(X-1). Thus ends the long gestation period of budget formulation, preparation, and review. The second major phase of the budget cycle, *congressional review*, begins.

The three stages of the congressional review phase, *budget, authorization,* and *appropriation,* begin in 198(X−2) and are conducted concurrently.The congressional budget stage begins when the Joint Economic Committee completes review of the five-year current services projects submitted by the president and reports its evaluation to the respective House and Senate Budget Committees. These committees begin consideration of the president's FY 198X budget by holding hearings in which agency heads, outside experts, and administration officials are questioned. In addition, the Budget Committees receive input from each of the Houses' standing committees (for example,

FEDERAL BUDGET CYCLE 131

House of Representatives' Committee on Science and Astronautics). The Congressional Budget Office, which has undertaken its own evaluation, also submits its report to the House and Senate Budget Committees. From these reports, the Budget Committees prepare the first FY 198X "concurrent resolution," which is submitted to the full legislative body (the House or Senate). This resolution defines total budgetary outlay for each revenue recommendation. Undoubtedly, major differences will exist between the House and Senate versions of the resolution. These must be ironed out in conference, after which the Budget Committees submit the second "concurrent resolution," which must be approved by both houses prior to the fall adjournment.

While these actions are taking place, the Legislative Committees of both houses take the FY 198X authorization bill, previously submitted with the 198(X−1) authorization bill, under consideration. The authorization bill also undergoes hearings, review, and debate until passed by each house. Resolution of Senate-House differences takes place in conference prior to final passage of the bill by both houses. The authorization bill becomes law only upon presidential signature or congressional override of the president's veto.

The *appropriations* stage begins in early 198(X−1), when the House and Senate Appropriations Committees begin their hearings on the FY 198X budget estimates. If the bill is approved or modified by the committees, it is then presented to their respective houses for debate. Again, differences in the House-Senate versions of the bill are resolved in conference prior to final passage. If the president vetoes the appropriations bill, Congress has the option of sustaining the veto or overriding it. If the former occurs, then the entire process moves back to the authorization stage and must be repeated.

Upon passage of an appropriations bill, the final phase of the budget cycle, *budget execution,* begins. The agencies develop an operating budget plan for FY 198X based upon congressional action on the appropriations bill. They then request apportionment of those appropriations by OMB. OMB reviews the apportionment requests with regard to the supporting operating budget plan. At this point, the agency's operating budget faces further perils, which will be discussed later. If no problems arise, the agency will *allocate* the funds available, under OMB-approved apportionments, to their operating subdivisions. These operating units then make decisions on use of the funds (grants, contracts, and so on) in accordance with agency procedures. Specific commitments, in accord with the planned expenditures, are approved, and the funds *obligated* upon execution of any binding document

Table 8-4. Example of a Detailed Budget for First Twelve-Month Budget Period (Personnel)

From: 04/01/X1   Through: 03/31/X2

Personnel Name/Position	%	Hrs. per Week	Salary	Fringe Benefits	Totals
Swinger, Grant, Ph.D., Principal Investigator	33		$14,850	$ 3,000	$17,850
Larry, I. R., M.S., Research Assistant	100		16,500	3,600	20,100
Mo, U. S., Research Technician	100		15,000	2,600	17,850
Curley, J. R., Animal Technician	50		7,000	1,200	8,200
Wheal, Daisy Clerk-Typist		20	7,280	1,250	8,530
Subtotals			$60,630	$11,650	$72,280

normally be those that the individuals currently receive, it is best to check with your personnel office to determine whether increases in wage scales during the next twelve to thirty-six months are anticipated. Keep in mind that nearly a year will pass from the time your grant application is submitted until the project start date. If a wage increase occurs during this period, and you are compelled to comply with the new scale, the costs burdened to the personnel category will be above budget by the time your grant becomes active. This means that you will either need to squeeze needed resources from other categories or seek supplemental funds—a rather inauspicious beginning.

After salaries have been listed, fringe benefits can be added. Fringe benefits are usually based upon the wage scale and, if you budget below the anticipated salary scale at start date, you will also be under-budgeted for this expenditure. In most cases your business office will be able to provide you with the dollar amount for fringe benefits associated with the particular salary. Often fringe benefits are based upon a percentage of payroll, but you may have to provide a description of just what benefits are included, depending upon the funding source. Among the typical fringe benefits often required are workers' compensation and unemployment insurance, Social Security (FICA), and vacation and sick leave. Depending upon your institution's policies, a range of voluntary benefits such as medical, life, and disability insurance and retirement programs may also be included. If your institution does not have a fringe benefits scale in dollar amounts for the particular salary listed in your grant you may then have to calculate the amounts for each person's benefits. For example: FICA at 7.12 percent × $14,850 (principal investigator's salary for 33 percent effort) = $1057.32; health insurance at $75 per month: 4 months × $75 = $300, and so on. This information may have to be broken down and listed separately anyway for some foundations.

Equipment. Most granting agencies allow for lease, rental, or purchase of equipment. Each item requested should be listed and fully described. If the equipment is to be leased, the terms must be provided, as shown in table 8-5. The NIH format requires justification for each item of equipment that is to be leased or purchased and the need for each must be apparent in the experimental protocol.

Supplies. Items of consumable supplies (materials to be used and expended in the project) must also be itemized by category. Unit prices and rate of consumption should be approximated. These costs must also be apparent to the reviewer. In the example, the research plan must call for 200 mice to be maintained for ten months as reflected in the budget.

Travel. The number of trips involved, the destination, the number

Table 8-5. Example of a Detailed Budget for First Twelve-Month Budget Period (General Expenditures)

Equipment (itemize)	
2 Hi-speed, refrigerated centrifuges @ $150/month lease × 12 months	$ 3,600
1 Gas liquid chromatograph and integrator	9,000
Supplies (itemize by category)	
Chemicals (biochemicals and solvents)	800
Glassware (cell culture flasks, disposable pipettes, test tubes, etc.)	1,000
Animals (200 mice @ $3.00/each)	600
Animal care costs (Per/diem) $.04/day; 200 × 300 days × .04	2,400
Travel	
Domestic: P.I. to meeting for presentations of project results, 4 days, 10th ASP meeting and to chair technical session	1,200
Foreign:	-0-
Other expenses (itemized by category)	
Reference Books	150
Reprographics, $30/month	360
Art and medical illustrations for publication	400
Page costs ($25/page) and reprint costs	600
Total direct costs	$92,390

Chapter 9
Contracts

The buying and selling of research through the vehicle of contractual arrangements has become an increasingly significant topic for the laboratory manager. While the federal government is an important source of contractual funding, many private research laboratories have studies conducted on a contractual basis. Both types of contractual studies will be discussed in this chapter.

CONTRACTS AND GRANTS

Contracts differ from grants in several ways. The most important of these is that a contract allows the contractor to add a fee to the basic costs of conducting the study. Other differences between grants and contracts are listed below:

- Any organization, profit or nonprofit, may bid on a contract.
- Most contracts are granted on a one-shot basis.
- The tasks to be completed under a contractual agreement are generally specified by the party awarding the contract, allowing very little flexibility to the contractor. A protocol for conducting the study is often made a part of the contract.
- Contracts are more closely monitored by the awarding agency, progress milestones are specified, and time schedules are emphasized.
- Interim reports may be specified to be submitted during the course of the study. Partial payments of the total costs are often based on the contractor's submission of these interim reports.
- Payments are normally made after the contractor has incurred the costs. However, in the case of some contractual studies, such

Donna A. Hurley

LABORATORY MANAGEMENT

LABORATORY MANAGEMENT PRINCIPLES AND PRACTICE

Homer S. Black, Ph.D.
Veterans Administration Medical Center
and
Baylor College of Medicine, Houston, Texas

Ronald C. Hart
Miles Laboratories, Inc.
Elkhart, Indiana

Orrin M. Peterson
Sandoz Crop Protection Corporation
Des Plaines, Illinois

Van Nostrand Reinhold Series in Managerial Skills in Engineering and Science

VNR VAN NOSTRAND REINHOLD COMPANY
NEW YORK

Copyright © 1988 by Van Nostrand Reinhold Company Inc.
Library of Congress Catalog Card Number 88-6
ISBN 0-442-21439-1

All rights reserved. No part of this work covered by the copyright hereon may be reproduced or used in any form or by any means—graphic, electronic, or mechanical, including photocopying, recording, taping, or information storage and retrieval systems—without written permission of the publisher.

Printed in The United States of America

Designed by Monika Grejniec

Van Nostrand Reinhold Company Inc.
115 Fifth Avenue
New York, New York 10003

Van Nostrand Reinhold Company Limited
Molly Millars Lane
Wokingham, Berkshire RG11 2PY, England

Van Nostrand Reinhold
480 La Trobe Street
Melbourne, Victoria 3000, Australia

Macmillan of Canada
Division of Canada Publishing Corporation
164 Commander Boulevard
Agincourt, Ontario M1S 3C7, Canada

16   15   14   13   12   11   10   9   8   7   6   5   4   3   2   1

Library of Congress Cataloging-in-Publication Data

Black, Homer S.
  Laboratory management: principles and practice/
Homer S. Black, Ronald C. Hart, Orrin M. Peterson.
    p. cm.—(Van Nostrand Reinhold series in
managerial skills in engineering and science)
  Includes bibliographies and index.
  ISBN 0-442-21439-1
  1. Medical laboratories—Management. 2. Laboratories—
Management. I. Hart, Ronald C. II. Peterson, Orrin M.
III. Title. IV. Series:
RB36.3.F55B58 1988
507'.24—dc19
                                                     88-6
                                                     CIP

Van Nostrand Reinhold Series in Managerial Skills in Engineering and Science

Michael K. Badawy, Series Editor
Virginia Polytechnic Institute and State University

Developing Managerial Skills in Engineers and Scientists: Succeeding as a Technical Manager by M. K. Badawy

Modern Management Techniques in Engineering and R & D by J. Balderston, P. Birnbaum, R. Goodman, and M. Stahl

Improving Office Operations: A Primer for Professionals by Jack Balderston

Managing the Engineering Design Function by Raymond J. Bronikowski

Applied Finance and Economic Analysis for Scientists and Engineers by James R. Couper and William H. Rader

Career Development for Engineers and Scientists: Organizational Programs and Individual Choices by Robert F. Morrison and Richard M. Vosburgh

Laboratory Management: Principles and Practice by Homer S. Black, Ronald C. Hart, and Orrin M. Peterson

Series Introduction

Laboratory Management: Principles and Practice is the seventh volume in the Van Nostrand Reinhold Series in Managerial Skills in Engineering and Science. The series will embody concise and practical treatments of specific topics within the broad area of engineering and R & D management. The primary aim of the series is to provide a set of principles, concepts, tools, and practical techniques for those wishing to enhance their managerial skills and potential.

The series will provide both practitioners and students with the information they must know and the skills they must acquire in order to sharpen their managerial performance and advance their careers. Authors contributing to the series are carefully selected for their experience and expertise. While series books will vary in subject matter as well as approach, one major feature will be common to all volumes: a blend of practical applications and hands-on techniques supported by sound research and relevant theory.

The target audience for the series includes engineers and scientists making the transition to management, technical managers and supervisors, upper-level executives and directors of engineering and R & D, corporate technical development managers and executives, continuing management education specialists, and students in technical management programs and related fields.

We hope that this dynamic series will help readers to become better managers and to lead most rewarding professional careers.

MICHAEL K. BADAWY
Series Editor

Preface

In preparing a work that spans a broad spectrum of knowledge and information, decisions concerning scope and content must be given considerable thought. Recognizing the practical impossibility of addressing the full range of laboratory and research management problems in a single volume, we have attempted to create a text that provides a solid foundation for the student and practicing manager, together with guidance in specialized areas that are of general concern. Our objectives are threefold. The first is to present the principles and concepts that constitute the science of management. It is upon these cornerstones that any approach to laboratory management must ultimately rest. Chapters on planning, implementation, and control introduce the basic elements of management, with emphasis on their application in the laboratory. Intended to serve as a primer, these chapters should prove particularly informative both to students and future laboratory managers following a scientific or technical career path.

Our second goal is to provide insight into some specialized aspects of laboratory management. Laboratory safety and effective technical communication are areas of particular importance in this field. The chapters treating these subjects should be helpful to professional managers who need to transfer their skills to the laboratory and research arena. The chapter on patents and the management of proprietary information will be of immediate interest to research managers whose laboratories are already involved in such activities. With more and more emphasis being placed on rapid transfer of technology, however, many other laboratory managers, including those in the nonprofit sector, will find this information beneficial.

Finally, we have attempted to offer a guide to the framework within which well over half of all laboratory managers must carry on their profession—the federal budget and regulatory systems. The chapter outlining the federal budget cycle provides the backdrop against which scientists and research managers engage in a never-ending quest for financial support of laboratory activities. Special emphasis is given to grants and contracts, the major instruments by which research is funded. The final chapter concerns regulatory constraints with which a laboratory must comply. It is not intended to be a complete and definitive reference source for this extremely fluid and changing area. Rather, our purpose is to make the laboratory manager aware of the many regulatory agencies, their major areas of regulation, and their impact upon how one goes about the business of managing a laboratory.

By presenting the cornerstones upon which management rests, but with emphasis on specialized aspects peculiar to the laboratory manager and within the operational framework in which most laboratory managers must function, it is our hope that this volume may serve as both a primer and a reference source for the effective practice of laboratory management.

Contents

**PART 1. PRINCIPLES AND PRACTICE OF
LABORATORY MANAGEMENT**   1

Chapter 1 Program Development and the Planning Process in the Laboratory   3
Missions   3
Long-Range Planning   4
Strategic Planning   5
Intermediate-Range Planning   10
Short Range Planning   12
Summary   18
References   18

Chapter 2 Program Implementation   21
Resource Management   21
Organizational Structure   26
Motivation of Personnel   28
Problem Solving and Decision Making   33
Summary   38
References   38

Chapter 3 Effective Control Techniques   41
Control of Research Endeavors—Meaning and Rationale   41
The Control Process   42
Specialized Objective Control Techniques   44
Financial Control Techniques   50
Other Fiscal Control Measures   51
Subjective Control Techniques   53
Technology Transfer and Project Abandonment   54

Summary 56
References 57

PART 2. SPECIALIZED ASPECTS OF LABORATORY MANAGEMENT 59

Chapter 4 Laboratory Safety 61
Management of Laboratory Safety 61
Laboratory Design and Equipment 70
Hazardous Materials 75
Special Topics 81
Summary 84
References 84
General Bibliography 85

Chapter 5 Effective Technical Communications 87
Seminars and Publications as an Incentive for Employees 88
Writing Research Reports 89
Publishing Research Results in Journals 89
Presenting Seminars 90
Graphics and Audiovisual Materials 92
Hints for Better Technical Writing 94
Communicating Technical Results to the Layman 95
The Obligation of the Technical Spokesman to the Public 96
Summary 98
References 98
General Bibliography 99

Chapter 6 Patents and Proprietary Information Management 101
Proprietary Information 101
Trade Secrets 102
Patents 104
Laboratory Notebooks 120
Licensing Technology 120
Inventorship Awards 121
Summary 121
References 122

PART 3. GOVERNMENT—ITS IMPACT ON RESEARCH MANAGEMENT 123

Chapter 7 Federal Budget Cycle: Impact on Science, Technology, and Research Management 125
Historical Perspective 125
Steps of the Budget Cycle 128

Impoundment, Continuing Resolutions, and Other Exigencies 132
Summary 135
References 136

Chapter 8 Grants and Grant Management 137
What Is a Grant? 137
Types of Grants 139
The First Steps in Applying for a Grant 141
Formal Proposal Preparation 145
The Review and Award Process 156
Grant Management 161
Summary 168
References 169
General Bibliography 171

Chapter 9 Contracts 173
Contracts and Grants 173
Contracting to Do Research 174
Sponsoring Contractual Research 178
Contractual Responsibilities 181
Summary 182
References 182

Chapter 10 Compliance with Governmental Regulations 183
Good Laboratory and Management Practices 183
Disposal of Hazardous Wastes 188
Pollution Control 192
Shipping Hazardous Materials 192
Recombinant DNA 194
Controlled Substances 199
The FDA and Food Additive or GRAS Approval 201
FDA Drug and Device Approvals 205
Equal Employment Opportunity and Affirmative Action 209
Summary 215
References 216

Appendix A Research Involving Human Subjects 217

Appendix B Research Involving Animals 227

Index 235

PART 1
PRINCIPLES AND PRACTICE OF LABORATORY MANAGEMENT

Chapter 1
Program Development and the Planning Process in the Laboratory

As a manager of a research unit, you should expect program development and planning to require an ever-increasing percentage of your time. A multitude of procedures go into putting together a comprehensive plan for the tasks your unit must carry out and the results that you and your subordinates need to achieve. However, prior to developing programs or implementing plans, you must have a clear understanding of your organization's mission.

MISSIONS

Every viable organization, whether it is a university, governmental group, or private industry laboratory, must have a mission. Definition and communication of this mission is the responsibility of all levels of management. In an organization whose management communicates effectively, the employee will have a clear, concise definition of that mission, whether the mission is to find a cure for cancer or to market and sell a product at a rate of return that is satisfactory to the stockholders.

Within each organization, the mission of the various departments comprising the organization will vary considerably. Generally speaking, R & D is concerned with finding and developing new products or processes, developing new applications for existing technologies, improving old products, or addressing contemporary problems of societal concern. The marketing department's mission is to develop marketing plans to introduce products to the market place, as well as developing test markets, advertising plans, and market research

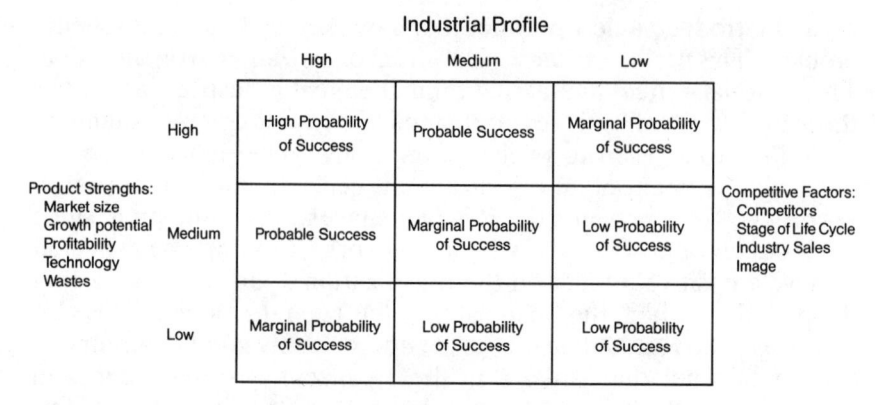

		Industrial Profile		
		High	Medium	Low
Product Strengths: Market size Growth potential Profitability Technology Wastes	High	High Probability of Success	Probable Success	Marginal Probability of Success
	Medium	Probable Success	Marginal Probability of Success	Low Probability of Success
	Low	Marginal Probability of Success	Low Probability of Success	Low Probability of Success

Competitive Factors:
Competitors
Stage of Life Cycle
Industry Sales
Image

Figure 1-2. Matrix analysis of a potential product or project.

Strategies that most commonly require input from the laboratory are those that involve

- Developing new products
- Maintaining the current position of a product
- Increasing sales of a product by either increasing the market share in current markets or by increasing product effectiveness
- Devising new uses for existing products
- Increasing profitability by decreasing manufacturing costs

After thorough analysis of all factors involved, the appropriate strategies are selected by management for individual products.

Since input from several areas is required to develop an effective strategic plan, overall responsibility for development of a specific strategy is generally assigned to a strategy manager. The area from within the organization from which the strategy manager is selected is determined by the particular strategy.

The strategy manager must accomplish several specific functions in order to develop a strategic plan for management review. These include

- Definition of assigned strategy
- Division of strategic effort into component programs
- Analysis of the costs and benefits expected if the strategy is accepted
- Establishment of critical checkpoints
- Measurement of performance in the execution of the selected strategy

In order to develop costs and benefits for a particular strategy, the strategy manager must obtain input from functional managers for such areas as research and development, marketing, manufacturing, regulatory compliance, environmental, and finance. These managers provide input on the factors involved in developing the strategy, including manpower, capital costs, working cash requirements, overhead, and general expenses.

Review and Modification of Strategies. After all recommendations have been made to the strategy manager, each proposal is reviewed to determine whether the risk involved in executing it will result in adequate benefits to the organization. At this point strategies are modified or alternative strategies are selected to replace those that are not compatible with the organization's objectives. An example of a strategy that might require modification is a new product development plan that requires several new products to be introduced in a short period of time—for example, within the following year. In order to accelerate development of these products, a substantial increase in R & D effort would be required. The increase in expense would be of such magnitude that profits for the organization would be reduced to a point below that set as a corporate objective. The strategy might be modified so that either a lesser number of new products is selected for development or the period over which the new products are to be introduced is extended. An alternative plan might involve acquiring "new" products—with most of the early phases of development completed—from another organization.

Plan Document. The output of the strategic planning process is a formal document detailing the strategies selected for a finite time period, typically between four and five years. Specific functions such as R & D, may extend their strategic plan for longer periods to allow the necessary time to develop new products.

When the plan document has been approved by the top management of the organization, it is usually published and distributed to all the functional managers affected. It is actually preferable to distribute a summarized version of the strategic plan to all employees, so that the strategies selected to guide the organization are fully understood by everyone concerned and each person understands his part in their successful execution.

Goals and Objectives

The use of goal-setting procedures has been in practice, on a formal basis, since the early 1960s. Informally, goals have probably been established for an even longer period. Management by Objectives (MBO), or Management by Results (MBR), is a formalized method for developing objectives that many believe leads to improved planning and organization of work. While it is admittedly more difficult to set objectives for technical personnel than for line or sales personnel, this system has been found applicable to the laboratory (Badawy 1976).

Objectives are best developed by a two-pronged approach in which general goals are transmitted from the top of the organization to the lowest level. Formalized objectives from the lower levels are then transmitted to successively higher levels of management, who modify their objectives to include those of the lower levels. This system results in organizational objectives that are a balance and blend of the individual goals of all levels of the organization.

In order for an MBO or MBR program to be successful, objectives must be

- Realistic
- Achievable
- Measurable
- Specific
- Consistent with responsibility
- Compatible with department and corporate objectives
- Understood by each level of management involved
- Revised upwards from one period to the next (increased effort should be required)
- Significant
- Presented in written form

McConkey (1972) states that "technical personnel generally require a larger number of more complex objectives than line personnel." A great deal of care must be taken, however, to ensure that numbers games are not played for the sake of writing measurable objectives. It is not enough, for example, for a senior scientist to state as an objective, "During the next year, I will synthesize twenty-four novel compounds." While this goal is measurable, it does not specify what purpose the compounds will have after synthesis. If your organization is geared toward developing new pharmaceuticals, synthesis of compounds designed to control the growth rate in crops or weeds may or may not be compatible with your organization's overall objectives.

Objectives should be reviewed periodically by both the writer and his superior to determine if progress is being made in achieving them. There may be valid reasons for modifying an objective during this review process, for example, if a decision has been made to decrease

R & D effort in a specific project area and to reallocate effort to other areas. Periodic review is a key concept in MBO systems. When objectives are inflexible, they tend to be less effectively written and do not provide a degree of stretch. The results of the reviews can and should be used as part of the individual's performance review. McConky (1967) suggests that compensation should be based on the degree to which an individual satisfies his or her objectives. In a situation where incentive compensation is applicable, the amount of incentive awarded would be based on the percentage of completion of objectives.

Management by Objectives is best implemented if the total organization is involved, from the top executive to the lowest echelon. However, an MBO program can be implemented on a departmental basis within an organization. The criteria for success of MBO, whether implemented in one department or in the entire organization, are identical. One of the most important of these is that the system must have the full and total commitment of top management, who must enthusiastically back the system, accept its ground rules, and actively participate in the goal-setting procedure. If MBO does not give the expected benefit, a thorough analysis is likely to show that failure was not due to inherent weaknesses in the system but rather to the way in which it was implemented. Management by Results is a refinement of the MBO procedures sometimes referred to as Management by Objectives for Results (MBOR). While both terms can be applicable to a specific plan, MBR tends to emphasize the shorter term, focusing on areas that are likely to lead to results in the least possible time.

Key Result Areas (KRAs). KRAs are those areas in which the manager can expect to achieve the maximum return for effort expended. Managers often find that the highest percentage of their time is expended on activities that yield the least in positive results. Some observers insist, with tongue in cheek, that we spend 90 percent of our time to achieve 10 percent of our results and 10 percent of our time to achieve 90 percent of our results. Morrisey (1977) suggests that a disproportionate amount of time is indeed spent on trivial matters and that by selecting the areas in which a greater output can be achieved with minimal input, the manager can operate more effectively. In a research environment, a typical manager may include the following in a list of key result areas:

- Innovation and imagination
- Product improvement
- Administration
- Budgeting

- Strategic planning
- Operational planning
- Staff development
- Self development

Suggested percentages of time to be devoted to each of these result areas cannot be projected because of varying levels of managerial expertise and the differing priorities associated with individual research managers' assignments. However, by determining the actual amounts of time required for each management area, a manager can assess his time utilization and redirect his efforts to areas that should be emphasized. When objectives are set, those related to key result areas should be given a higher priority than less productive objectives.

INTERMEDIATE-RANGE PLANNING

Project Selection

Selection of projects to be funded in a research program is a key consideration because of the need for a profitable outcome and the fact that unlimited funding is extremely rare. Industrial research funding is generally limited to a set percentage of the sales of the corporation, a level that varies widely depending on the type of business in which the organization is engaged. The range may extend from as low as 1 percent of sales to as high as 15 to 20 percent in some of the high-technology industries. On average, corporations spend slightly over 3 percent of their sales dollars on R & D (Gerstenfeld 1970). The chemical industry's average spending on R & D is between 3 and 5 percent of sales (Webber 1985). Because of limited funds, a procedure must be established to ensure that precious research dollars are allocated to those areas that will bring the maximum benefit to the organization with the lowest risk of failure. In assessing benefit-versus-risk factors, the following questions need to be considered:

- Is there a need for the product or process?
- Is this project compatible with our current business?
- Can adequate patent or trade secret protection be obtained on this product?
- Will the product or process be environmentally acceptable?
- Are hazardous wastes generated?
- Can the product be produced to sell for a price that will give an adequate return on investment?

- Will this product lead into new areas for the organization? If so, is this desirable?
- How much capital will be required to build or modify a plant for production of this product?
- Does the project have societal value?

In most organizations some type of a project proposal that details the answers to these questions is required. Its form may vary considerably from one organization to another. One approach is shown in table 1-1.

Project Evaluation

In industry, an effort is made to direct funding to those projects that will have the greatest desirable impact on the immediate success of the organization. The criteria used to determine which projects will be funded may vary from one organization to another depending on strategies chosen by the particular organization. However, most evaluations will include the following:

- Resources required (including internal manpower, outside contractual resources)
- Projected time required to obtain the desired results
- Probability of success

Table 1-1. Research Project Proposal

From:_____ Date_____

Project Title:_____

  A. Objective
  B. Project description
  C. Plan of action
  D. Justification
      1. Economic factors
      2. Competitive factors
      3. Technical factors
      4. Market potential
  E. Resource requirements
      1. Manpower
      2. Time
      3. Capital equipment

- Projected return on resource investment
- Environmental factors

Within a balanced research program with adequate funding, there should be a mix of projects to provide a new product pipeline that will support the organization in the future. Short-term projects with a high probability of success and potential for increased corporate profits will receive funding priority. However, longer-range projects with payback periods of increasing length need to be included in the R & D program to provide products and processes to be introduced in the future.

SHORT-RANGE PLANNING

Since most organizations operate on a yearly budget, a short-range or operating plan must be developed. It should be consistent with the long-range objectives of the organization and supportive of the departmental and individual objectives that have been developed. Individual managers within the R & D organization must develop and monitor an operating plan that meshes with the overall objectives of the department. Long-range implications also need to be considered. If a long-range project is to be started within the operating plan period, it may affect succeeding operating plans for a considerable period after the original plan has been completed. Many chronic toxicology studies, for example, require three to four years to complete, with costs exceeding $500,000. A variety of project management techniques may be used to develop an operating plan.

Networking

PERT/CPM (Program Evaluation and Review Technique/Critical Path Method) networks are extremely useful in creating an operating plan. In a well-developed network, events to be completed within a given time frame can be readily identified and estimates of the time and dollar resources required can be developed. The groups responsible for the completion of each event are also clearly identified. The most important feature of the network, however, is the establishment of interrelationships between start and completion of individual events and identification of events dependent upon other events for initiation or completion. Those events in which a delay in completion would result in a delay of the entire project are identified as being on the critical path of the project. In developing a PERT/CPM network, a completed event is signified by a circle or oval. The activity required to complete that event is represented by the line connecting two events, as illustrated in figure 1-3. Some component events do not require completion of other such events prior to their initiation. These are called independent events, as shown in figure 1-4. The time or cost requirements required to complete the event can be shown on the activity line. The activity lines can also be made proportional to the time or cost requirements for completion of an event (fig. 1-5). The critical path signifies those events that, if delayed, would cause a significant delay in completion of the project. Figure 1-6 shows multiple pathways that must be completed prior to completion of the project. In this case, pathway 1-3-5 is the critical path.

Gantt Techniques

Gantt techniques often provide a better representation of the time periods required to complete individual tasks within a project. Events

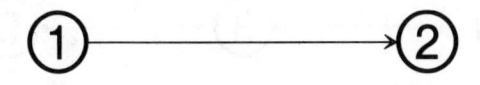

Events
1. Start construction of pilot plant
2. Complete construction of pilot plant

Activities
1-2. Construct pilot plant

Figure 1-3. Networking.

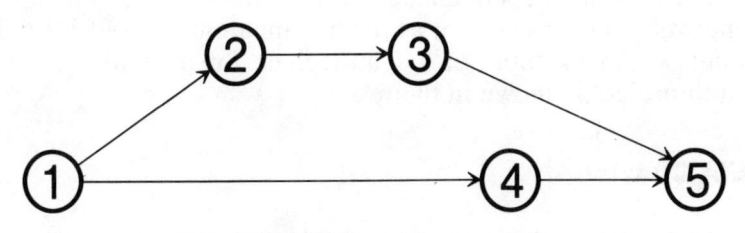

Event 1. Construction started
Event 2. Footings and foundation poured
Event 3. Roof completed
Event 4. Landscaping completed
Event 5. Pilot plant occupied

Figure 1-4. Independent events.

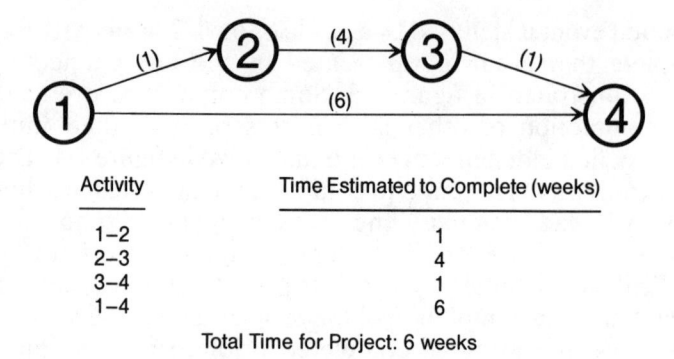

14   PRINCIPLES AND PRACTICE OF LABORATORY MANAGEMENT

Activity	Time Estimated to Complete (weeks)
1–2	1
2–3	4
3–4	1
1–4	6

Total Time for Project: 6 weeks

Figure 1-5. Time relationships.

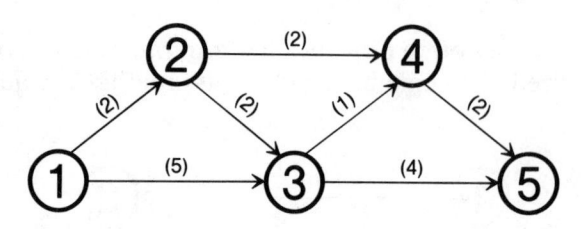

Figure 1-6. Critical path. The 1-3-5 path is the critical path; the rest are slack paths.

directly dependent on the completion of another event are shown sequentially on the same line of the chart. However, when initiation of a particular event is dependent upon the completion of several events, the interrelationships are not as readily apparent as in PERT/CPM techniques. An example of a Gantt chart for a relatively simple research project is shown in figure 1-7.

DELTA Charts

In most projects, decision points must be programmed to allow for adoption of alternative courses of action in case the original action plan is found wanting. Use of a DELTA technique allows these decisions to be preprogrammed and to present alternative courses of action on the original plan. An example of a typical DELTA chart is shown in figure 1-8.

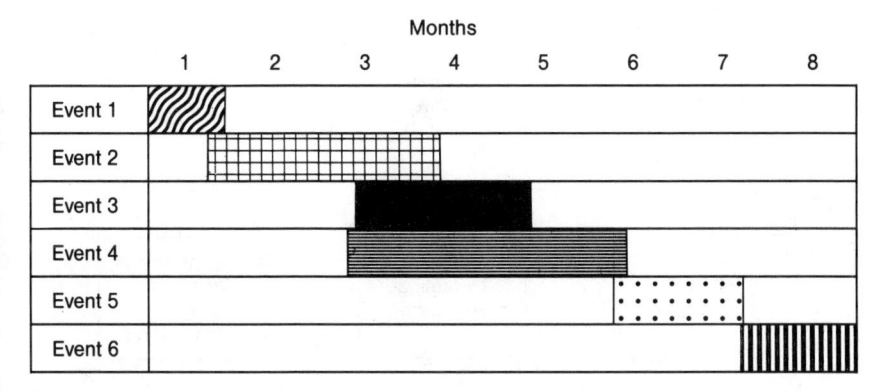

Figure 1-7. Gantt chart.

Computerized Techniques

All of the above project management techniques are amenable to computerization. Use of the computer to determine the effect of alternate courses of action or the effects of a delay in completion of tasks along the critical path can give management an important decision-making tool. Costs of implementing alternative courses of action are easily developed and revised as necessary. However, the software package used to handle project management tasks should be selected cautiously. Many software packages are available for different hardware configurations. Some of these packages are more suitable for engineering-type projects than for chemical laboratory projects. Make sure that the program will give you the type of information you require.

Budgeting

Each project plan should be accompanied by an individual project budget. The sum of all project budgets will comprise the organization's annual budget. This raises an interesting question: which should come first, the plan or the budget? The answer is only slightly less complex than the answer to the original question regarding chickens and eggs.

In developing a project budget, it is feasible to use current-year figures for research costs as a starting place, adjusting these figures to account for external factors such as inflation. In 1984, the average cost per industrial R & D scientist or engineer was $127,000 (anon.

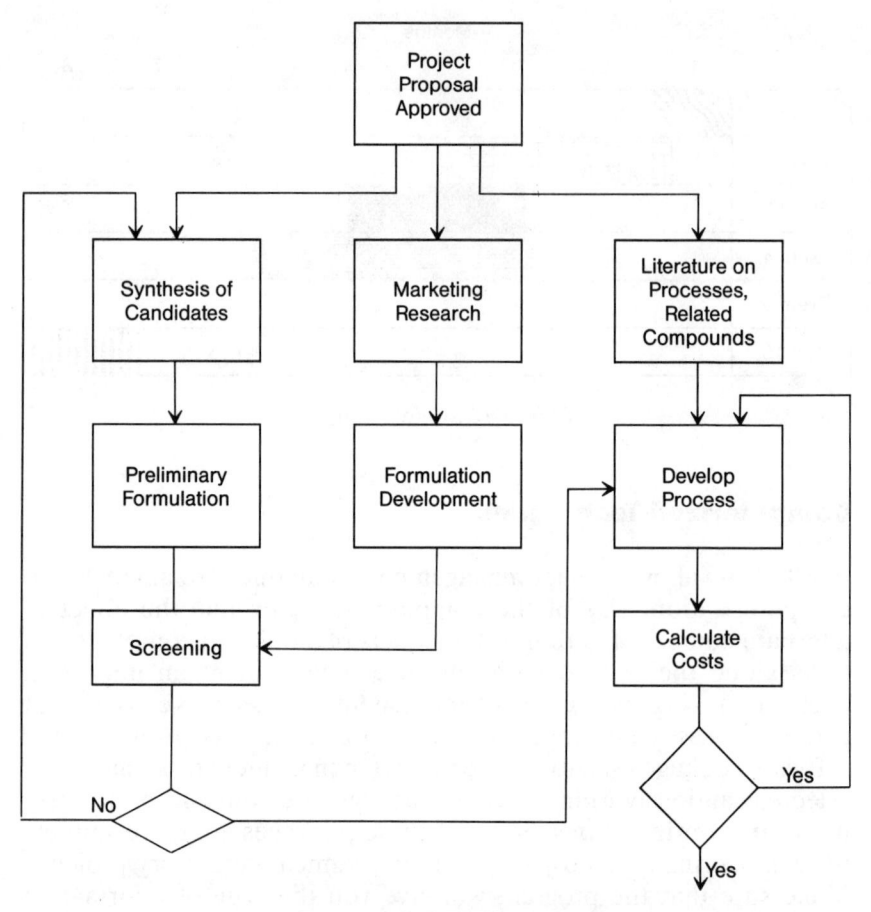

Figure 1-8. DELTA chart.

1986). An example of development of project manpower cost is shown in table 1-2.

Using current figures, individual project budgets can be projected based on the manager's estimate of the manpower required to achieve the set objectives. A sample project budget is shown in table 1-3. Cost projections for each individual project must take into account the input and requirements of all departments within the organization. This stage of the plan development program requires extremely good communication among the various departmental representatives to obtain meaningful projections of requirements.

Table 1-2. Project Labor Charge Development

Total R & D budget 1985	$12,000,000
Cost of outside contractual work	$ 3,000,000
Net internal R & D costs	$ 9,000,000
Number of scientists employed	75
Cost per man-year	$ 120,000
Projected 1986 labor costs (@ 5% annual inflation)	$ 126,000

Table 1-3. Project Budget: Product Technical Support

	Projected Man-Years	Labor & Overhead	Outside Contractual	Total
Technical Services	3	192,600	0	192,600
Analytical	4	256,800	10,000	266,800
Formulations	2	128,400	50,000	178,400
Process Development	1.5	96,300	20,000	116,300
Pilot Plant	3	193,600	30,000	222,600
Total	16.5	866,700	110,000	976,700

Capital Requirements

Purchase of equipment required to support individual projects may be handled separately from the expense budget. For example, if a new gas chromatograph is required to support a particular project, its costs must be included in the total cost of the project but can be placed in a separate capital budget with equipment needed for general support or replacement. A capital purchases plan is an essential part of a well-developed operating plan. Good financial management also requires that the finance group be aware of the requirements to purchase expensive capital equipment so that funds can be made available without having to resort to expensive short-term borrowing. Conversely, the finance department should be notified if purchase of equipment is to be delayed so that funds can be diverted to other uses or be invested in appropriate short-term vehicles.

Changes to the Plan

After management has allocated funds for R & D for an operational plan period, changes may be required to modify, cancel, or add projects. If good project selection and evaluation techniques have been used in developing the plan, this task becomes infinitely easier. However, projects that have strong "champions" may be more difficult to modify than those that are not as strongly supported, regardless of the level of funding.

The plan may be subjected to drastic changes during the course of its execution. Most researchers have had the experience of being asked to "put out a fire." The reason may be a request from a regulatory agency to provide additional information regarding product safety, field reports that a product is not performing according to design specifications, or the introduction of a new product by a competitor with superior properties to that being sold by your company. Each case offers sufficient justification for diversion of R & D effort from planned projects to the "firefighting" project. Management must be advised of the effect of any diversion of resources on the overall research program. In many cases a decision based on an either/or situation must be reached. Without an additional infusion of resources, it is not possible to accomplish both "firefighting" and the originally established program. Fortunately, some "firefighting" projects can be completed in a relatively short time by temporary reassignment of personnel from other projects. Cost increases caused by lack of continuity on the planned projects are extremely difficult to estimate. They are usually allocated as part of the cost of completing the "firefighting" project.

SUMMARY

In this chapter we have tried to present the general relationships between long-range planning, strategic planning, organizational and individual objectives, and budgeting—both expense and capital. A planning cycle (fig. 1-1) can be established that includes each of these interrelated functions.

REFERENCES

Anon. 1986. "Facts and Figures for Chemical R & D." *Chemical and Engineering News* 64:32-60.
Badawy, M. K. 1976. "Applying Management by Objectives to R & D Labs." *Research Management* 19:35-40.
Gerstenfeld, A. 1970. *Effective Management of Research and Development.* Reading, Mass.: Addison-Wesley.

McConkey, D. D. 1967. *How to Manage by Results.* Rev. ed. New York: American Management Association.
──────. 1972. "Staff Objectives Are Different." *Personnel Journal* 51:477-83.
Morrisey, G. L. 1977. *Management by Objectives and Results for Business and Industry.* 2d ed. Reading, Mass.: Addison-Wesley.
Webber, D. 1985. "Chemical Industry Will Spend 9% More on R & D in 1985." *Chemical and Engineering News* 63:19-20.
Weil, E. D., and R. R. Cangemi. 1983. "Linking Long-Range Research to Strategic Planning." *Research Management* 26:32-39.

Other Suggested Reading

Badawy, M. K. 1983. "Why Managers Fail." *Research Management* 26:26-31.
Geigold, W. C. 1982. *Practical Management Skills for Scientists and Engineers.* Belmont, Cal.: Lifetime Learning Publications, 232-34.
Gibson, J. E. 1981. *Managing Research and Development.* New York: John Wiley.
Merrifield, D. B. 1978. "How to Select Successful R & D Projects." *Management Review* 67:25-39.

Chapter 2
Program Implementation

When an acceptable operational plan has been developed for the laboratory, it is the manager's responsibility to carry out the activities required to reach the objectives of the plan. A manager must consider several very important areas to obtain the results needed to satisfy his objectives.

RESOURCE MANAGEMENT

Management has been defined as "coordination and integration of resources to accomplish specific results" (Scanlan 1974). These resources include money, people, and facilities. The manner in which a manager uses the resources allocated to him can have the greatest impact on progress in solving the particular set of problems assigned to him, as well as affecting the degree to which he advances in his career.

Funding/Financial Resources

All research managers, whether in the profit or not-for-profit area, face a common problem regarding funding. Allocation of research funds to those areas that will provide the most significant results is essential to a successful research program. The manager must know the amounts spent to support each of the projects assigned to him and their potential for giving a positive result.

As a manager, you will have a specific amount of funds allocated for use by your laboratory. Continued support of your work either through grant renewal, new grants, or support of your management in allocating R & D funds to your laboratory depends in part on how well you manage the funds available to you.

In this age of computerization, most organizations have established procedures to account for the expenditures of each area of responsibility within the organization. Monthly or weekly reports of charges to your responsibility area can easily be generated and matched against the amounts budgeted for your research.

Direct and Indirect Costs. Two types of costs must be considered. The first are those under your direct control, known as controllable costs. The second are costs over which you have no direct control, sometimes called departmental or sectional overhead. The latter costs are generally added to the actual costs generated by your area by your accounting department to provide accurate reporting of the total costs for each project being conducted within the organization. Examples of typical direct and indirect costs are shown in table 2-1.

Each reporting period gives you an opportunity to review the amounts spent in support of the projects under your control as well as the total amounts spent against the funds budgeted for this work. Two examples of expense spending reports are shown in tables 2-2 and 2-3.

Time Reporting. The key to obtaining an accurate estimate of the costs incurred for each project assigned to your laboratory is a realistic time-reporting system. This system requires that the individual researchers accurately report the amount of time spent on each of their projects. While a detailed minute-by-minute report is seldom needed, the individuals should try to conform within the degree of accuracy you need to manage your limited financial resources efficiently.

In some cases of decentralized management, you may be required to develop your own automated procedure for comparing project costs with budgeted costs. Many microcomputers can be programmed to reduce the amount of time required to determine how funds are being spent and to indicate, based on project progress, any possible reallocations of funds (Nixon 1983).

Table 2-1. Direct and Indirect R & D Costs

Direct Costs	Indirect Costs
Salaries	Related payroll costs
Materials/supplies	Occupancy charges
Travel	Allocated support
Contractual costs	Administration
	Clerical

Human Resources

Allocation of the human resources available to you may be the most challenging part of your managerial role. As with funds, the number of personnel available to you may be severely limited. You may also be restricted to personnel whom you did not personally select. While large organizations may have a pool of manpower available to be reassigned for specific projects, you may be faced with having to assign existing personnel to specific project assignments. Several factors must be considered in allocating manpower effectively.

Your estimate of the manpower requirements for each project should be incorporated into an operating plan. This plan should

Table 2-2. Sample Expense-spending Report

MONTHLY EXPENDITURES, SEPTEMBER 1984
ORGANIC SYNTHESIS GROUP

	Actual	Budgeted	Variance
Salaries	21,893	22,000	107
Project supplies	2,376	2,000	<376>
Glassware apparatus	1,123	1,200	77
Chemicals	956	1,000	44
Travel	376	500	124
Telephone	22	20	< 2>
Related payroll	3,284	3,300	16
Total	30,030	30,020	< 10>

Table 2-3. Sample Expense-spending Report

PROJECT STATUS REPORT

Research Manager: *J. Smith*   *May*

Project No.	Man Months	Salaries	Project Supplies	Overhead	Total
1	0.52	3,100	150	1,040	4,290
2	1.55	9,300	950	3,100	13,350
3	1.45	8,750	1,056	2,900	12,706
4	0.68	4,027	220	1,360	5,607
Total		25,177	2,376	8,400	35,953

support the strategic plan objectives set for your area and be supported by the funds budgeted for your area and allocated to specific projects. The operating plan specifies how many people will be required to complete the needed work in a given period of time. In some cases, you may be able to increase the number of people assigned to a project in order to complete the work in a shorter period. In other cases, it may be counterproductive to assign additional personnel to a project. As you gain in experience and can judge the personnel reporting to you based on past performance, the art of manpower allocation becomes easier. Initially, you may have to rely on your own judgment based on your personal experience in a "bench" (non-supervisory) situation.

In selecting personnel, you need to make sure that the proper people are assigned to the project. Among the factors that enter into such decisions are the work experience and particular strengths of the candidates. You should be familiar with the background of each of the people reporting to you, to aid in the selection of the most appropriate individuals for a project. Increasingly, organizations in both the public and private sectors require that a curriculum vitae be on file for all technical personnel. This information, which may be required to support a grant application, respond to a request for a proposal, or satisfy regulatory agency requirements, can help in identifying the most suitable personnel.

The personal characteristics of each of the people reporting to you must be considered in assigning projects. Factors you may wish to consider include

- *Work habits:* Does he work better alone or in a group situation?
- *Reaction to pressure:* How does she react to establishment of stringent deadlines?
- *Degree of supervision required:* Can he take over on a project and run with it, or will he require a considerable amount of close supervision?

These factors can have a significant impact on allocation of your personal time, affecting the amount of attention you can devote to other important aspects of your work as well as the success of the project itself.

Outside Contractual Services

After you have evaluated the personnel available, you may find that your group is weak in one or more specific areas. In this case an chemists, chemical engineers, and analytical personnel. An organization-wide project team might be established for the introduction of a new product. In this case the team might be composed of representatives of sales, marketing, product development, research, manufacturing, legal, and finance. Another term for the use of project teams is matrix management.

Matrix Management. For large, complex projects that require technical input from diverse groups of personnel, a team system known as matrix management may be used. The matrix is composed of as many people as are needed from each line organization to achieve the project objectives. These personnel report technically to a project manager and administratively to a functional manager. Good matrix management is characterized by

- Good project definition
- Definite start and definite stop
- A single focused job to do

Matrix management can lead to both real and imagined conflicts between project managers and functional managers. The degree of conflict can be expanded if all projects in the program are designated as matrix projects. If functional managers have personnel participating in several matrix projects, the potential for conflict with each of the project managers is magnified.

Matrix management has been used successfully in large-scale projects requiring diverse technical input (Epperly 1981). An example of the organizational structure of a matrix project team is shown in figure 2-1.

Task Groups

Task groups are most likely to be composed of "birds of a feather" —that is, people having similar backgrounds. A specific task is assigned to the group and all members contribute to solution of the problem.

The Individual Researcher

In some cases, the most appropriate organizational structure will be a group of individual researchers, each assigned to specific projects within your area of responsibility. This situation is most likely to

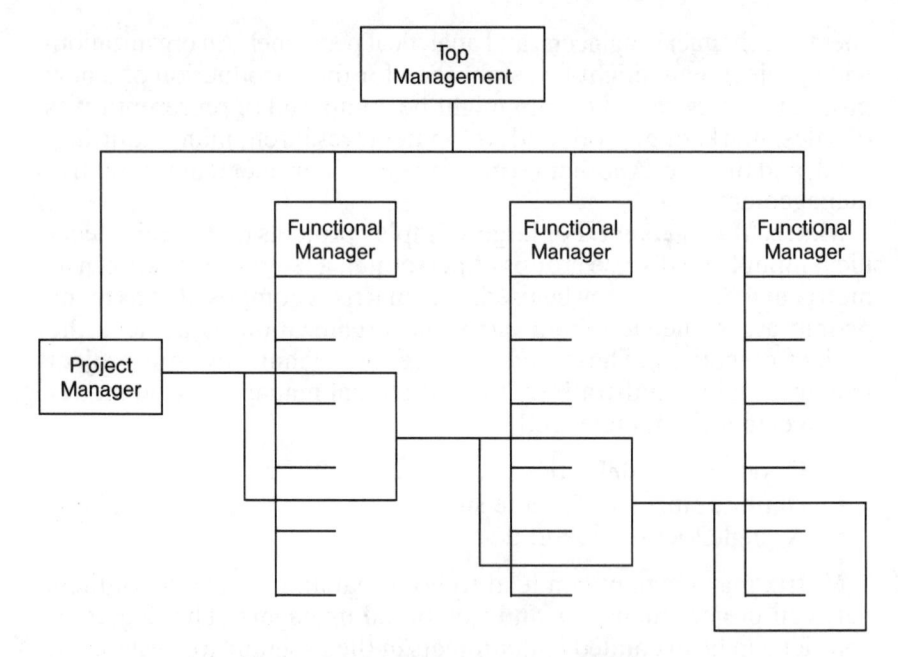

Figure 2-1. Matrix management team.

be found in smaller research organizations. (The term *individual researcher* applies to a scientist or engineer supported by technicians.)

MOTIVATION OF PERSONNEL

Motivation is a complex phenomenon. You as a manager cannot motivate the personnel reporting to you, as motivation is an individual's response to a set of internal needs and external stimuli. You as a manager can have a significant effect on the motivation of personnel by providing them with the particular external stimuli that will cause them to be motivated.

The scientists and engineers in your organization were selected to fulfill a creative function. This fact sets them apart from production workers or most service industry personnel. Providing motivation for creative personnel is in itself one of the most difficult challenges to the R & D manager. Wortman (1983) defines important traits of the creative person as "(a) the courage to be different and (b) a dedication to long hours and hard work."

Motivational Theory

In order for you to provide an environment that will motivate your personnel, you must have a basic understanding of the theories of motivation and you must know each individual well enough to make certain that the specific factors that will provide motivation for him are present. There are several theories as to what provides motivation. Although each has strong points, you may need to select and merge elements from different theories to provide adequate motivational forces.

Maslow's Theory. Maslow (1954) describes a hierarchy of needs that reveals itself as a person matures in everyday life and in the work environment. The factors that motivate the chairman of the board of a corporation that exceeds a billion dollars per year in sales are certainly not the same as those that motivate the B.S. chemist fresh out of college. Maslow's rule of thumb is that the lower-order needs must be satisfied before any of the higher-order needs can be fully addressed. Although his hierarchy has usually been illustrated by the use of a pyramid (fig. 2-2), it may also be represented by a series of steps (fig. 2-3).

Herzberg's Theory. Herzberg, Mausner, and Snyderman (1959) observe that two sets of factors comprise each job situation. The first set, which he calls maintenance or hygiene factors, must be satisfied before any motivational influences can be operative. (These are very similar to the lower-order needs in Maslow's hierarchy.) Herzberg lists a set of motivational factors that he believes are required to provide an atmosphere conducive to obtaining the best performance (table 2-4).

McGregor Theory X, Theory Y. How managers deal with motivation was explored by Douglas McGregor (1960), who proposed that managers could be divided into two categories. The Theory X manager basically believes that all employees are lazy and dislike work. The Theory X manager uses fear and punishment as his primary motivational factors—especially fear of loss of the job and fear of criticism in front of peers. (In the early industrial revolution in America, the majority of managers subscribed to the Theory X tenets.) On the other hand, Theory Y managers assume that most people will respond to motivators and produce high quality work because it gives them a sense of accomplishment.

Most managers will need to use a combination of management techniques at one time or another in their management careers. The need to change from one style to another can be from day to day or hour to hour depending on the personnel involved. While it is likely

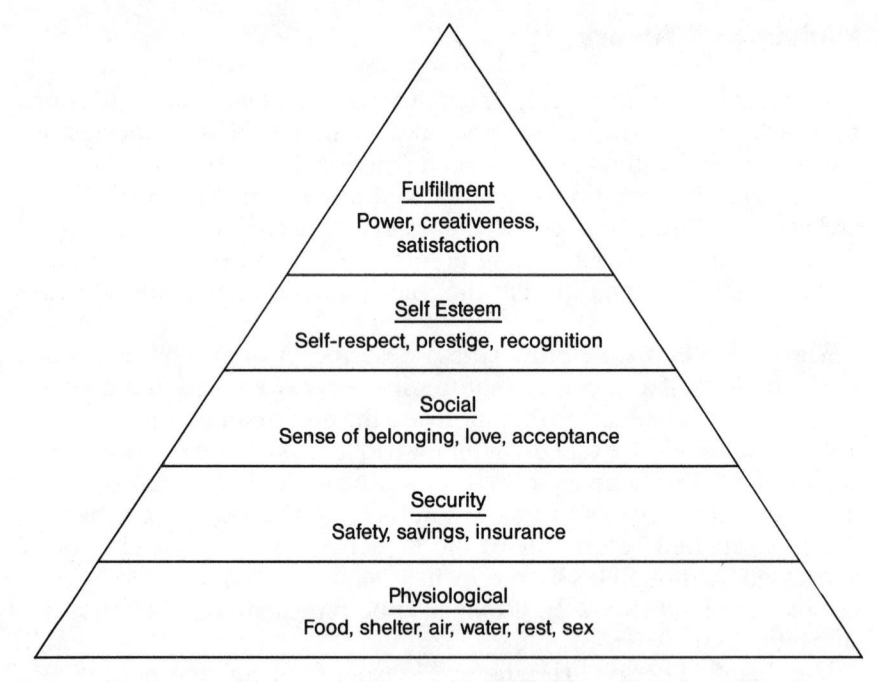

30   PRINCIPLES AND PRACTICE OF LABORATORY MANAGEMENT

Fulfillment
Power, creativeness, satisfaction

Self Esteem
Self-respect, prestige, recognition

Social
Sense of belonging, love, acceptance

Security
Safety, savings, insurance

Physiological
Food, shelter, air, water, rest, sex

Figure 2-2. Maslow's hierarchy of needs—Pyramid Diagram (Maslow 1954).

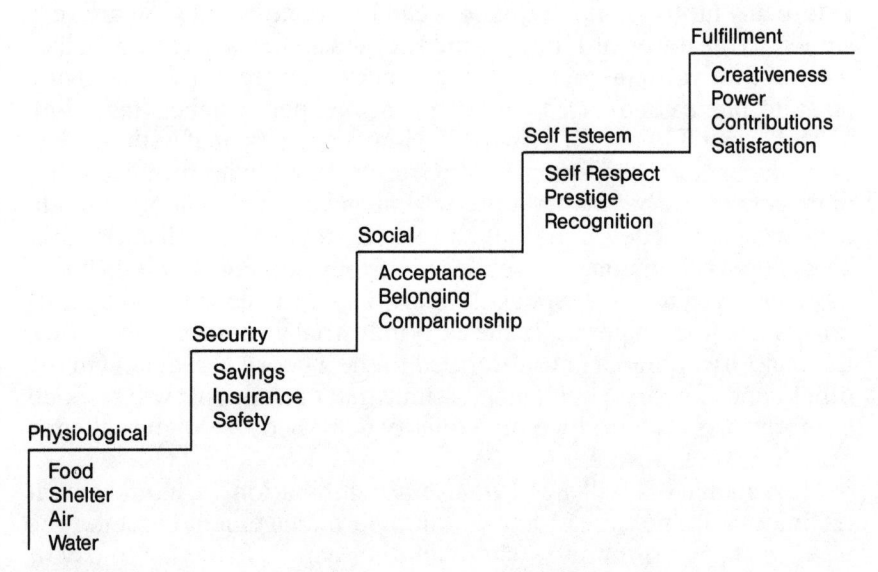

Fulfillment
Creativeness
Power
Contributions
Satisfaction

Self Esteem
Self Respect
Prestige
Recognition

Social
Acceptance
Belonging
Companionship

Security
Savings
Insurance
Safety

Physiological
Food
Shelter
Air
Water

Figure 2-3. Maslow's hierarchy of needs—Step Diagram (Maslow 1954).

Table 2-4. Herzberg's Factors

Maintenance Factors	Motivational Factors
Adequate salary	Achievement
Working conditions	Recognition
Job security	Responsibility
Company policy	Advancement
Technical supervision	Job satisfaction
Interpersonal relationships	

SOURCE: Herzberg, Mausner, and Snyderman, 1959.

Table 2-5. Styles of Management

Theory X	Theory Y
Dictatorial	Participative
Structured	Supportive
Driven	Motivational
Conservative	Empathetic

SOURCE: McGregor, 1960.

that few managers could be classified exclusively as either Theory X or Theory Y types, many managers tend to use the techniques of one more than the other. Within these two management styles, McGregor distinguishes a number of variations, which are listed in table 2-5.

Position Design

In order to provide the appropriate motivational factors for each of your employees, you need to consider several factors in designing individual position descriptions. Most personnel will function more effectively in a position that has been designed to take into account their strengths and to minimize areas in which they are weak. A job can rarely be designed to match a person's current abilities completely, but potential for acquisition of needed abilities can be used to make the match more satisfactory. The following factors should be considered in designing positions within your area of management.

Authority and Responsibility. How much authority will each employee have in conducting his assignments? This may include responsibility for the personnel reporting to him, the amount of funds that he may commit without further approval, or even specification of the direction in which a project proceeds. You should report back to the employee on the effect of his use of the authority you delegated to him.

Matching People with Tasks. As a manager, one of your most important duties is to be able to evaluate each employee with respect to his individual abilities and to match the assignments to his abilities. For example, people who do not enjoy direct supervision of others should be assigned those tasks that do not emphasize this function.

Growth. Determine the areas in which each employee needs to grow and attempt to provide levels in your organizational structure into which he can reasonably expect to move, given satisfactory acquisition of new skills and knowledge.

Individuals in the Laboratory

Within any laboratory you are likely to find a wide variety of people, each with individual traits and characteristics. Remember, you selected them for their creative abilities, and creativity may be masked under many cloaks. Some of the types of individuals you are likely to encounter are discussed in the following paragraphs.

The Team Member. This person works well within a team effort. He makes substantive contributions and interfaces well with those in the group. His suggestions are usually well thought out and respected by those within the group. This person should be given an opportunity to assume more managerial duties.

The Loner. This person prefers to be given an assignment and to be allowed to carry it out on an individual basis. He may be just as creative as any of your personnel but prefers to keep to himself. He may resent constant supervision and suggestions.

The "Prima Donna." Every laboratory will have at least one "prima donna." Problems arise, however, when your work force is composed of several "prima donnas," each of whom demands that his needs be constantly catered to. This demand diminishes the amount of time you have available to tend to the motivational needs of your other employees.

The Old Pro. This type of employee has been around long enough to know what is required of him to retain his job and accomplishes only enough to keep it. Generally he fits under the Theory X manager's concept of a typical employee.

The Follower. The employee who carries out instructions well but does not exhibit a high degree of innovativeness is usually the backbone of your organization. Caution is required, because he can do considerable damage to your program by doing exactly what you told him to do, and this may be his intent. By lack of common sense or lack of initiative, this person could act detrimentally to the organization by adhering to the letter of the instructions rather than using common sense or good judgment. At times, damage to the project could be intentional.

These are the most common types of workers who inhabit the laboratories of the world, though others can be identified. Your challenge as a manager is to recognize the type of individual with whom you are dealing and to adjust your management techniques to present each individual with a challenging set of motivators.

Recognition

Perhaps one of the strongest motivators for professional staff is recognition by management and peers of substantive accomplishment. Many organizations recognize accomplishment by awarding titles. An announcement on the bulletin board that Dr. X has been promoted from Assistant Research Associate to Research Associate, in recognition of his work in developing a new process, may have more lasting motivational influence than the raise that accompanied the promotion. (The salary increase will of course have some impact.) Some organizations select a Researcher of the Year, publicly presenting a certificate or plaque that may or may not be accompanied by a monetary award (Turner 1979).

PROBLEM SOLVING AND DECISION MAKING

Problem Solving

Problem solving is a multi-step procedure that can be divided into two processes: (1) analyzing the problem and (2) choosing one of the alternate solutions that have been identified (Leavitt 1972). Most technical personnel are constantly involved in the problem-solving process. Their effectiveness to the organization and to you as manager is highly dependent upon how well they understand the procedures for solving problems. Many technicians approach a problem intuitively with a high degree of success, while others require considerable guidance. Your performance as a manager will be evaluated on how well you solve not only problems that involve technical matters but also those that are concerned with people, budgets, and schedules.

Kinds of Problems. A problem may be defined as a situation or event that does not conform to your plans or expectations (Kepner and Tregoe 1965). Many problems in everyday life are the result of unexpected happenings. An automobile accident may happen because a driver flashing his left directional signal fails to make the expected turn. The driver's last minute decision to turn right creates a problem for you. In your laboratory you expect a subordinate to complete an assigned task by the deadline you have agreed upon. If the task is not completed on time, problems may be created.

Problems may be generated by factors that you can control. They may also be generated by factors over which you have no control, illustrating Murphy's Law and its multitude of correlaries. Many problems are created by a misunderstanding of objectives. When you make an assignment to one of your subordinates, it is much easier to make sure that he thoroughly understands his specific objectives prior to starting the task, rather than to try to solve problems created by his misunderstanding after the fact. When setting task objectives you should be as specific as possible and the authority required for completing the task should be clearly defined. If the person assigned to the task does not have authority to initiate a purchase of capital equipment, for example, the objective-setting process should specify the person he should rely upon either to obtain the needed authority or to obtain the needed action from his superiors.

In some cases a problem is created by an error. It is your job to determine how the error was made and to implement procedures to minimize the likelihood of similar errors being repeated in the future.

Problem Definition. The first step in the problem-solving process is to define the problem by obtaining answers to the following questions:

- What happened?
- What should have happened?
- What is the difference between what was expected to happen and the actual result?
- Is it really a problem?
- How bad is the problem?

The answers to these questions will help you in determining a course of action to correct the problem. Avoid attempting to solve problems that in reality do not exist. To quote an old saying, "If it ain't broke, don't fix it." The next step in the process is to separate symptoms of the problem from the actual cause of the problem.

Finding the Solution. Once you have separated symptoms from causes, you should list all possible solutions to the problems and their ramifications. You are then in a position to decide upon a satisfactory course of action. However, prior to making that decision you may wish to determine how the problem could have been prevented in the first place. Identify all possible causes of the problem and any actions on your part or on the part of those reporting to you that would have prevented the problem from arising. Past experience should provide insight on how to handle a similar situation in the future.

Decision Making

In your management career, you will be called upon to make decisions as a part of your daily duties. Ranging from trivial to major, these decisions will influence the degree of success or failure of your unit. As a part of the planning process, you are asked to select the optimal alternatives to achieve your objectives. In problem solving you are required to select a workable solution from the possible alternatives. The alternatives selected may not be the best of those identified, but should be the best compromise available under your current circumstances. The most effective manager is the one whose decisions are based upon sound judgment and are made in a reasonable time period.

Decision-making Styles. Just as managers vary in the type of management style they employ, the way they handle decision making can also be characterized. Giegold (1982) lists five types of decision makers: the totally autocratic manager; the less autocratic manager; the consultive manager; the more consultive manager; and the consensus manager, who decides only on the basis of a consensus of his subordinates. The logical decision maker, who gathers as many facts as possible and evaluates alternatives, generally—but not always—comes up with fair and sound decisions.

Decision-making Criteria. There are several criteria involved with decision making. The first of these is that the decision should be made at the lowest possible level at which all the data necessary for the proper action come together (Bowden and Miller 1979). In other words, the decision should be made by the person who is closest to the action. As decisions are made at levels further and further from where the action is required, the "facts" become more and more complex and confusing. The "facts" to the chairman of the board of a corporation are often not those that are apparent to a research manager.

A second criterion is that once a decision has been made, it should be communicated clearly to those who will be involved in the actions required. A third is that the decision maker must follow through to determine if the actions required by his decision are in fact being carried out.

Another important guideline is to retain flexibility, leaving yourself room to pull back if at some point you find that you did not select the best alternative. This is not to imply that you should vacillate in making decisions. However, there will be times in your career when you will be forced to make decisions without an opportunity to evaluate all the alternatives or obtain all the pertinent facts.

Decision-making Aids. A wide variety of decision-making guidelines to aid the research manager have been proposed in the management literature. (See, for example, Jackson 1983, Merrifield 1978, and Foster 1971.) Among the areas addressed in detail is the determination of which projects are to be funded within the R & D program. New projects should be proposed with a view to (1) their compatability with the current mission of the organization, (2) the amount of resources required to conduct the research, and (3) the ultimate effect of doing the work, with respect to the organization supporting the work, the enrichment of scientific knowledge, or the benefit to the general public. It is extremely difficult to forecast the benefit of a project that may be completed seven to ten years from its initiation, and even near-term forecasting is a complex process.

Checklists or profile charts using weighted criteria are often used

Table 2-6. Project Profile

Department	Concerns	Rating[a]
Research	Time required	2
	Personnel	4
	Chance of success	1
Marketing	Potential markets	3
	Competition	4
Manufacturing	Raw materials	1
	Equipment	2
	Cost	1
Legal	Patent/license situation	4
Corporate	Fit to current business	1
	opportunity for new technology	4
Average Rating		2.45

[a]Ratings: Unfavorable, 1-2; Neutral, 3; Favorable, 4-5

to determine which new projects are to be initiated. Examples of profile charts are illustrated in tables 2-6 and 2-7. Obviously, the criteria and the weight assigned to each criterion must be based on the current strategies being employed. Another method of selecting a new project is to develop a decision tree, using a basic logic sequence that asks three questions (Merrifield 1978):

1. Is this a good opportunity for anyone to choose?
2. Is this a good opportunity for us to choose?
3. If yes, what is the best way for us to get involved?

An example of this type of decision tree is shown in figure 2-4. Other types of decision trees require an estimate of the probability of success of each of the steps involved in the completion of a project (Jackson 1983).

Various methods are used to determine whether to accelerate, slow down, or terminate development on current projects; to calculate cash flow and internal rate of return; develop benefit-to-cost ratios. None of these methods can accurately assess the benefit of a development such as the Salk vaccine or a herpes antigen. Nor can they predict the serendipitous type of event that has occurred time after time in the guise of a breakthrough. Decision-making aids that focus on financial analysis satisfy the accountants to some extent, but their use by laboratory managers on a regular basis is not well documented (Foster 1971).

At times it will be necessary to use one or a combination of decision-making methods. Other circumstances will require you to make a "best shot" decision without using a formal method.

Table 2-7. Project Scoring

Criteria	Project Score[a]	Criterion Weight	Weighted Score[b]
Cost	6	10	60
Probability of success	3	8	24
Marketability	4	8	32
Development time	3	6	18
Fit to business	4	4	16
Total			150

[a]Scoring scale: Excellent = 10, Poor = 0
[b]Maximum weighted score: 500

```
         ┌──→ No ──────→ ┌─────────┐ ←──────────────────┐
         │               │ ABANDON │                    │
┌────────┴────┐          └─────────┘                    │
│ Is this a good │                                      │
│ opportunity to │                                      │
│   be in?       │                                      │
└────────┬───────┘        ┌────────────────────┐        │
         └──→ Yes ──────→ │ Is this a good opportunity │──→ No
                          │  for our organization?     │
                          └────────┬───────────────────┘
                                   │              ┌──────────────┐
                                   └── Yes ─────→ │ What is the best │
                                                  │   way to get     │
                                                  │    into it?      │
                                                  └──────────────────┘
                                                    Internal R & D
                                                    Acquisition
                                                    Joint venture
                                                    Licensing
                                                    Other
```

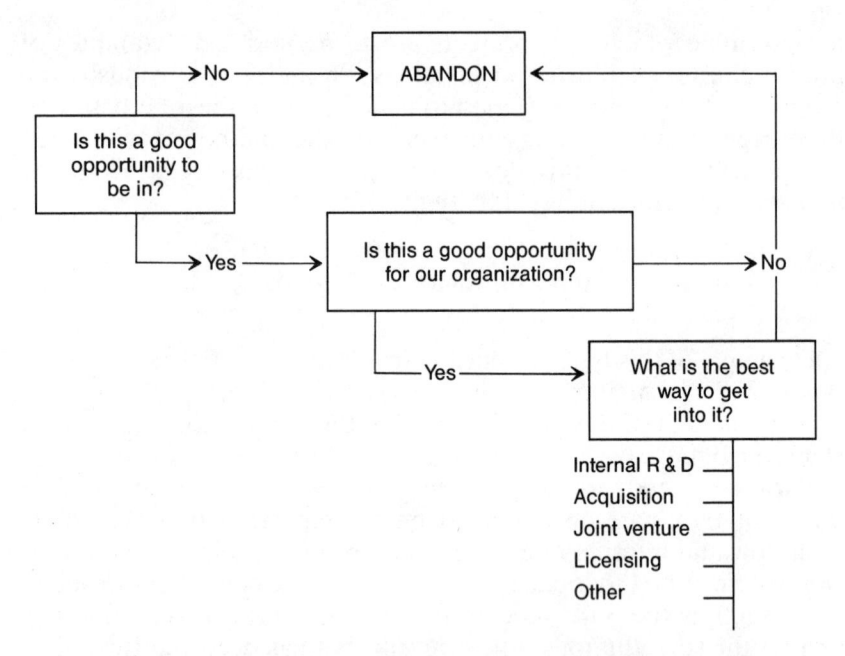

Figure 2-4. Basic logic sequence. (Courtesy D. B. Merrifield, 1978, "How to Select Successful R & D Projects," *Management Review* 67:25-39)

SUMMARY

In this chapter we have tried to give you a background in some of the most basic skills you will require in a management career. It is not enough, however, to learn these principles; they need to be reviewed and modified in light of your ongoing experience. The science of managing R & D is far less exact than the science practiced in your laboratories. You should also be aware of new theories of management developed for other areas of your organization that may have an impact on the management of your laboratory.

Periodic review of how you practice these management principles, as well as continuing education by means of seminars, short courses, and keeping abreast of the literature, is also required if you are to remain both technically and managerially competent.

REFERENCES

Bowden, J., and H. Miller. 1979. *The Effective Manager.* Naperville, Ill.: Deltak Inc.
Epperly, W. R. 1981. "Matrix Management Can Work." *Chemtech* 11:664-67.

Foster, R. N. 1971. "Estimating Research Payoff by the Internal Rate of Return Method." *Research Management* 14:27-43.
Giegold, W. C. 1982. *Practical Management Skills for Engineers and Scientists.* Belmont, Cal.: Lifetime Learning Publications.
Herzberg, F., B. Mausner, and B. B. Snyderman. 1959. *The Motivation to Work.* 2d ed. New York: John Wiley.
Jackson, B. 1983. "Decision Methods for Evaluating R & D Projects." *Research Management* 26:16-22.
Kepner, C., and B. Tregoe. 1965. *The Rational Manager.* New York: McGraw Hill.
Leavitt, H. J. 1972. *Managerial Psychology.* 3d ed. Chicago: University of Chicago Press.
Maslow, A. H. 1954. *Motivation and Personality.* New York: Harper and Brothers.
McGregor, D. 1960. *The Human Side of Enterprise.* New York: McGraw Hill.
Merrifield, D. B. 1978. "How to Select Successful R & D Projects." *Management Review* 67:25-39.
Nixon, R. A. 1983. "BUGPROG: An Automated Budget Preparation Procedures." *Journal of the Society of Research Administrators* 14:41-46.
Scanlan, B. K. 1974. *Management 18: A Short Course For Managers.* New York: John Wiley.
Turner, W. J. 1979. "How the IBM Awards Program Works." *Research Management* 22:24-27.
Wortman, L. A. 1983. *Effective Management for Engineers and Scientists.* New York: John Wiley.

Chapter 3
Effective Control Techniques

CONTROL OF RESEARCH ENDEAVORS— MEANING AND RATIONALE

From the standpoint of definition, little clarification or improvement can be made upon Henri Fayol's comments of nearly seventy years ago concerning the control process:

> The control of an undertaking consists of seeing that everything is being carried out in accordance with the plan which has been adopted, the orders which have been given, and the principles which have been laid down. Its object is to point out mistakes in order that they may be rectified and prevented from occurring again [Hodgetts 1979].

It becomes obvious from this statement that the principles of control are dependent upon the planning process discussed in chapter 1, for without a clearly defined plan and objective there can be no effective control.

One may argue that the events manifest of a research or creative undertaking cannot be as highly structured as that of a manufacturing and marketing plan in which accountability is reflected in the bottom line of a financial statement or annual report. By the same token, however, the very nature of scientific method demands that any research project follow a definite protocol or plan. Thus even in academia and its environs the basic parameters (a research plan) for controlling research activities exist.

The nature of modern research, with its high risk of failure coupled with increasing complexity and costs, has accentuated the need for effective control measures. Costs in particular have played an important part in the application of conventional financial control measures to research efforts. Burton Dean, in an extensive study published in 1968, found that companies in the United States at that time realized an identifiable return of only 45 percent on their R & D investment dollar. Thus, if a company realized a 10 percent rate of return from sales of a newly developed product, that company would have to sell over half a million dollars worth of the product for each $100,000 invested in its development, just to cover the costs of unproductive research efforts. Based upon an econometric model, Braunstein et al. reported in 1980 that most research was profitable for private firms, although the median private rate of return was only 25 percent. The median social rate of return from research investments was estimated at 56 percent. One can readily see that innovative research endeavors have at least a 50 percent chance of failure. Thus, from either a societal or proprietary point of view, the selection and effective management of research efforts becomes of great economic importance. It seems obvious that those involved in the management of research must utilize the methods and techniques employed in the management of any high-risk, limited-resource venture if the optimal return on the research dollar is to be realized. Specifically then, the control of a research effort refers to the methods and procedures used to regulate the project's progress and performance and to contain its costs.

THE CONTROL PROCESS

Managerial control is analogous to the task of the navigator, who must periodically take a positional fix and make the appropriate corrections to bring the craft back on course. Whether a laboratory falls short or exceeds its planned objectives, revisions to that plan will be necessitated, and this is where effective control measures come into play. In its basic elements, the control process consists of

- The delineation of objectives or aims as set forth in the research plan
- Comparison of performance against those objectives
- Correction of variations from the plan that may have occurred

The first element, delineation of objectives, should always be an integral part of planning and therefore information of which the laboratory manager is acutely cognizant. Understand, however, that predetermined objectives are not etched in stone and that in complex and lengthy endeavors the objectives must at times be realistically modified. Thus, standards of performance based upon original objectives will only be useful in the initial stages of any project that later undergoes modification.

In order to achieve the second element, there are certain requisite conditions to be fulfilled. First, there must be methods of *evaluation* by which the status of the project can be assessed. This is essential if one is to pinpoint variations and their cause(s). Second, an effective *feedback* system needs to be established. The requirements for these prerequisites are similar: they both must be expedient, timely, flexible, and economic (Hodgetts 1979).

Expediency: Evaluation and feedback systems should provide only information that is needed to control the research activity. The laboratory manager must determine what information is essential. Because the primary goal of any information system is to provide decision-making information, the laboratory manager should be responsible for design of both evaluation and feedback systems. In this way he is assured that essential information is obtained in an understandable and useful fashion. Remember, a deluge of unnecessary and superfluous data can be as detrimental as too little information.

Timeliness: The control measures implemented by the laboratory manager should quickly flag performance variance. Fayol, in his definition of control, states that "the object is to point out mistakes in order that they may be rectified and prevented from occurring again." Indeed, most managers practice control by the exception principle; that is, they address those situations that are especially bad or good. However, a well-designed control system should include lead factors that indicate the potential for variance before it occurs.

Flexibility: As noted earlier, control measures for some projects will be suitable for only the early or initial phases of the project. If those control measures indicate alteration of objectives, then the control measures themselves will undoubtedly need modification to comply with the newly established goals. Thus control measures must be flexible enough to be effective during periods of transition to alternate objectives and amenable to modification themselves.

Economic viability: The methods of control to be employed must be worth the expenses in time and money. A small laboratory, for instance, would not normally require, nor could it afford, elaborate computerized program evaluation and review techniques. On the other hand, projects requiring the coordination of work from several laboratories, perhaps some remotely removed from the evaluation site, might find the cost-benefit ratio of such an elaborate control measure not only favorable but the implementation of such a measure essential. Even the need for another progress report should be closely scrutinized with respect to economy. Scientists nowadays spend a great deal of their time preparing research protocols and grant applications. Additional demands upon their time should be carefully justified.

SPECIALIZED OBJECTIVE CONTROL TECHNIQUES

Many of the tools in planning projects discussed in chapter 1 are also necessary constituents of the control process. Those planning techniques provide the manager with an overview of the project, a method of determining how different facets of a project are interrelated, a means of identifying potential problem areas, and a measure by which overall progress can be assessed. These attributes are also essential ingredients for control. The most successful applications to the control process involve some form of time-event analysis.

Gantt Chart

An early (1917) technique of time-event analysis, named after its developer Henry Gantt, is still in widespread use today. The Gantt chart is a visual method for relating activities to performance times. This technique provides a two dimensional analysis with the horizontal axis depicting time—that is, work scheduled and completed—and the vertical axis identifying individuals and/or processes or procedures.

In figure 3-1, one can see that three types of experiments are being conducted during a one-month period. The cell culture and biochemical studies are labor intensive and the labor burden falls during the first three-week period. Thus, with limited technical assistance the manager may wish to control this situation by initiating the animal studies (2) in the second week—thus spreading the labor-intensive operations more evenly over the entire month. A dashed line has been used to depict the progress of each project. As seen at the end of week 3, project 1 is on schedule; project 2, having been delayed a week due to the revised schedule is now due for completion during week 4 and is on schedule; and project 3 was completed on schedule.

Milestone

An interesting modification of the Gantt chart that has gained popularity where scheduling and control procedures are required for less complex projects is *milestone* scheduling or analysis. This technique employs bar graphs to monitor project progress. The procedure

EFFECTIVE CONTROL TECHNIQUES   45

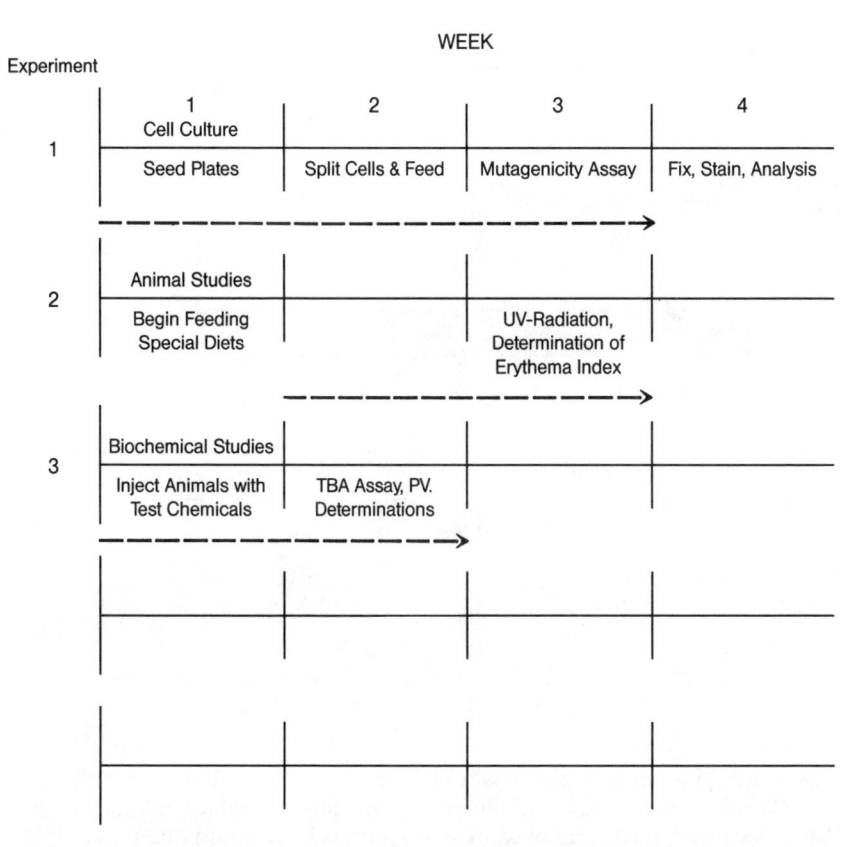

Experiment	WEEK			
	1	2	3	4
1	Cell Culture			
	Seed Plates	Split Cells & Feed	Mutagenicity Assay	Fix, Stain, Analysis
2	Animal Studies			
	Begin Feeding Special Diets		UV-Radiation, Determination of Erythema Index	
3	Biochemical Studies			
	Inject Animals with Test Chemicals	TBA Assay, PV. Determinations		

Figure 3-1. Gantt chart for laboratory operations. Activities are represented on the ordinate as experiments (1) cell culture, (2) animal studies, and (3) biochemical studies. The time frame for startup, progress, and completion of each activity is represented on the abscissa.

has been used successfully by several federal agencies where more sophisticated undertakings are winding-down and where a simpler control technique can be employed more economically. An example of milestone analysis is shown in figure 3-2. Our previous three projects, cell culture, biochemical analysis, and animal studies are depicted as milestone bars 1, 2, and 3 respectively. In this case the cell culture studies were begun at the beginning of time period 1 and are scheduled for completion at the end of time period 5. They are on schedule at the end of time period 4. The biochemical analyses were begun at the beginning of time period 2 and scheduled for completion

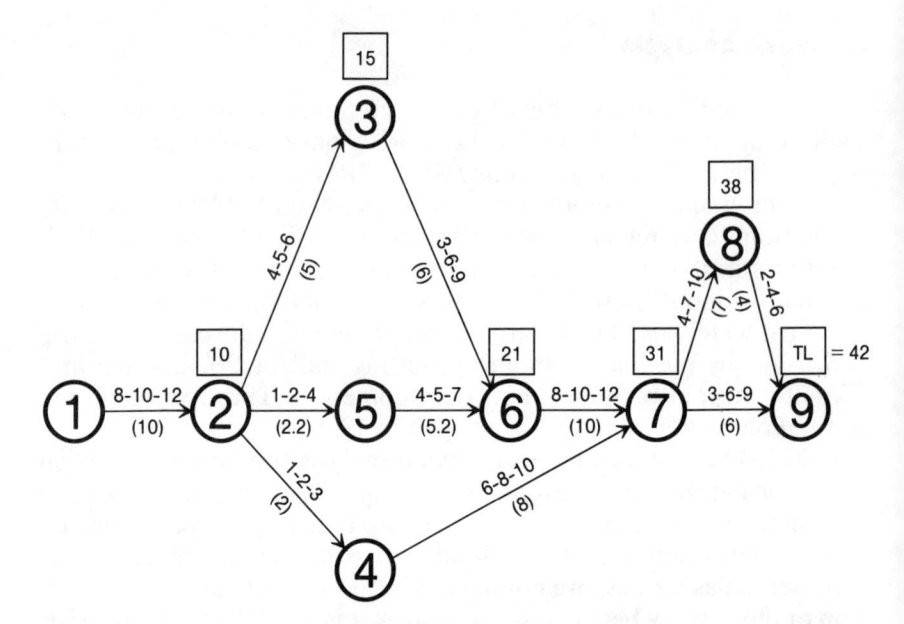

Figure 3-3. PERT flow chart. Milestones are designated by circles: numbers above arrows (8-10-12 and so forth) represent the most optimistic, T_O; likely, T_M; and pessimistic, T_P, times for completion of each respective milestone. Numbers in brackets below the arrows represent the actual computed expected time, T_E, for completion. The numbers in boxes above the milestones designate the critical path and are summed to reflect the earliest projected time to project completion, T_L.

projected time to completion is designated T_L. For our project in figure 3-3, the $T_L = 42$ weeks. *Slack* is the time difference between scheduled completion calculated for the critical path and the sum of the T_Es for each of the other paths. Slack can also be calculated for completion of each individual event.

$$\text{SLACK} = T_L - T_E$$

Table 3-1 represents the various parameters necessary for PERT analysis.

When making decisions concerning situations governing the progress of a project, the manager is faced with both risk and uncertainty. Risk is often defined as those situations in which the probabilities of a particular occurrence (success or failure) are known, whereas with uncertainty the probabilities are unknown. What we need to help us

Table 3-1. PERT Analysis: Parameters required for analysis of activities represented in figure 3-3

No.	Path	T_E (Weeks)	T_L	Slack	σ^2
1	1-2-3-6-7-8-9	10-5-6-10-7-4	42	0	2.68
2	1-2-5-6-7-8-9	10-2.2-5.2-10-7-4	38.4	3.6	2.82
3	1-2-3-6-7-9	10-5-6-10-6	37	5	2.24
4	1-2-5-6-7-9	10-2.2-5.2-10-6	33.4	8.6	2.38
5	1-2-4-7-8-9	10-2-8-7-4	31	11	2.43
6	1-2-4-7-9	10-2-8-6	26	16	1.99

in our decision-making situations then is a measure of risk that provides information about the tightness of our projected time estimates around a probability distribution—that is, the probability that our estimates are accurate. Determination of the variance of each event in our network is a useful statistical parameter that aids in our decisions and can be easily derived from the following formula:

$$\sigma^2 = [(T_P - T_O)/6]^2$$

As indicated by McLoughlin (1970), the divisor 6 is an empirical value that provides an adjustment for data distributions most likely to occur in a network. The variance not only provides us with an estimator of accuracy for each projected event time, but the variance of the total network, obtained by determining the largest sum of variances along any path through the network, allows calculation of the standard deviation (σ). This parameter is then used to establish the probability of completing the program according to the previously calculated T_L. For example, we find in table 3-1 that the sum of variances along path number 2, 1-2-5-6-7-8-9, is 2.82—the total network variance. Thus the standard deviation is found by extracting the square root of 2.82:

$$\sqrt{2.82} = 1.68$$

With a normal distribution, we know that the probabilities that the actual completion time will fall within 1, 2, or 3 standard deviations on either side of our critical path (T_L) are 0.683, 0.954, and 0.997 respectively. Thus, for our example we have a 68 percent probability that the project completion time will fall between 40.3 and 43.7 weeks.

As noted by Francis (1977), network analysis methods are best suited for project management or nonrecurring undertakings rather than continuous or ongoing operations. For obvious reasons they are applied most economically and efficiently to large, complex projects. Although these methods do provide a basis for identifying and correcting critical areas, they have been criticized for not taking into account the possibility that a network activity might fail or might not even be required to reach established goals. The technique of *network switching* was developed to interpose a measure of logic to the strongly time-oriented control method. Network switching involves the introduction of decision boxes at critical points in the network. At the decision points, selections can be made between alternative pathways based upon their probabilities of success. Network switching adds the dimension of logic, as well as time expectancy for project completion, to the control parameters available to the research manager. This leaves network analysis with one major shortcoming—inadequate emphasis on cost. Recently, the development of PERT/COST has addressed the question of financial considerations as a control measure.

FINANCIAL CONTROL TECHNIQUES

To establish fiscal control of a research project, it is essential that the budget be directly tied to technical progress estimates. Lin and Vasarhelyi (1980) have succinctly defined fiscal control in the answer to this question: "How is the project progressing in relation to budget revenues and expenditures?" To establish this relationship, the basic network analysis has been modified (using PERT/COST) to provide a method of placing the anticipated total cost of a project on the same statistical control basis as its technological progress. As fiscal control is primarily one of controlling expenditures, PERT/COST modification provides the manager with a budget standard with which he can control projected budgets against unplanned overruns.

The total costs associated with a complex project are for the most part direct costs consisting of salaries, materials, and equipment. To simplify the cost-technical progress relationship, Dean (1968) has assumed that total costs can be expressed as a linear function of direct labor costs—the latter being easily summed taskwise. We are now able to estimate future costs as a parameter having the same statistical properties and handled in the same fashion as our previous time-event analysis. PERT/COST relationships are represented in this simple diagram and equation:

$$\underset{\text{(1)}}{\textcircled{1}} \longrightarrow \underset{(t)}{\textcircled{2}} \longrightarrow ij \longrightarrow \underset{(t)}{\textcircled{3}} \longrightarrow ij \longrightarrow \underset{T}{\textcircled{4}}$$

$$COST_{(S)} = COST_1(t) = COST_2(t)$$

where $COST_{(S)}$ is the sum or total cost; $COST_1(t)$ is the actual cost incurred to date (say from event 1 to 2 in our diagram); and $COST_2(t)$ is the estimated future costs to time T (events 2-3 and 3-4).

We can now estimate future costs if we allow $cost_{ij}$ (cij) to represent costs/unit time associated with each activity (ij) and by using our previous three time estimates, that is, T_{Oij} (optimistic), T_{Mij} (most likely), and T_{Pij} (pessimistic). If we let t_{ij} equal the actual time attributable to a given activity (ij), then $COST_2(t)$ equals $\Sigma_i \Sigma_j c_{ij} t_{ij}$. The approximate mean cost ($\mu [t]$) for a particular activity is derived as

$$\mu(t) = 1/6 \, \Sigma_i \Sigma_j \, cij \, (T_{Oij} + 4T_{Mij} + T_{Pij})$$

The mean of the total direct cost $M[COST_{(S)}] = COST_1(t) + \mu(t)$. Probabilities of the accuracy of our cost estimates can be handled in a similar manner as for time-event analysis. With these methods the manager is able to estimate project total direct cost ($M[COST_{(S)}]$) each time a PERT reporting period occurs. Remedial action can then be implemented when future cost estimates exceed those budgeted. One should remember that in this simplified example only direct labor costs were considered. However, it has been estimated that one-half to three-fourths of major R & D expenses occur in personnel and thus some justification exists for this simplified approach. For most complex projects, direct costs including materials and equipment, as well as indirect costs chargeable to the project, must be considered. For the most part, these are costs over which the manager can exert some degree of control.

OTHER FISCAL CONTROL MEASURES

One of the most widely used long-term fiscal control measures in business is the financial statement. However, serious questions, as a result of applied accounting principles, have historically diminished the usefulness of the financial statement as a control measure for R & D. Proprietary organizations have usually charged R & D expenditures to current expenses rather than deferring or capitalizing an appropriate part of them. This situation has developed primarily from a standardized, conservative approach taken by accounting organizations such as the Financial Accounting Standards Board and as a consequence of both accountants' and managements' inability to assess expected future benefits of R & D projects. As noted by Lin and Vasarhelyi (1980), many firms' long-term success and profitability depend upon the continuing development of new and competitive products. Whereas the management of a firm may improve short-run profits through the application of imprecise and inflexible accounting principles, in the long run the dysfunctional impact of such treatment of R & D expenditures will eventually be felt—both in terms of an organization's profitability and the nation's R & D infrastructure. Even short-term effects on internal corporate decision making could be distorted due to inaccurate accounting practices regarding R & D expenditures and benefits. Recently, tax laws for R & D expenditures have partially eased this problem, while not addressing it directly. The financial statement remains an inaccurate measure of potential benefits derivable from specific projects and should be regarded accordingly when used as a control measure.

The same inadequacies reflected in the financial statement—the inability to identify and assess future benefits of individual research projects—limit the value for R & D of other fiscal control measures used in production and marketing operations. Nevertheless, some mention should be made of at least two methods that have proved useful to certain proprietary research operations, particularly the high tech and pharmaceutical industries.

Return on Investment (ROI)

ROI analysis measures how well an organization is performing with the assets it has at its disposal; that is, it answers the question "How well are we doing with what we have?" For example, if Company A had total investments of one million dollars in their research operations and realized $200,000 profit, while Company B had only $5,000 invested with $6,000 profit, there can be little doubt about which company is most profitable in terms of gross revenue. However, when examined on the basis of ROI, Company B far outperformed Company A in regard to how efficiently the investment dollar was utilized. The major shortcomings of this control measure are: (1) while useful to the company, it is difficult to apply in a meaningful manner at the laboratory level, and (2) without applying it in conjunction with other control measures, too much emphasis may be placed on fiscal considerations.

Break-even Point (BEP)

For a proprietary laboratory to be profitable to the parent company, revenues generated from the products or services provided by that laboratory must exceed total costs required to support the laboratory's operations. Conventionally, costs are classified as *fixed* or *variable*—based upon their behavior in relation to the level or volume of activity.

Under most management structures, fixed costs are those that remain constant, regardless of activity level. Examples would include equipment maintenance, service contracts, and labor. Variable costs are those that vary directly and proportionally with the level of activity. Expendable items used in research represent typical variable expense.

To compute the *BEP,* three cost/revenue parameters must be considered: total fixed cost (*TFC*), variable cost per unit of product or service (*VC*), and selling price per unit of product or service (*P*). The relationship is expressed in the formula:

$$BEP = TFC/(P - VC)$$

If, for example, a proprietary clinical laboratory estimated that they could service the medical community of a large metropolitan area with 20,000 special chemistry tests of a particular type at $25 per test, their total fixed costs set at $100,000, and variable costs per test $5, then,

$$BEP = \$100,000/(\$25 - \$5) = 5,000$$

The laboratory would have to market 5,000 tests before breaking even. If they approach their projected market of 20,000 tests, the venture would indeed yield a handsome profit.

One can see from the preceding formula that BEP emphasizes the margin-above-cost concept where variable costs are subtracted from selling price. This is an effective control technique, since any change in the variable or fixed costs will immediately affect BEP. Management therefore has a handle by which individual variables can be identified and appropriate corrective action can be taken.

The two preceding methods of fiscal control are obviously applicable at the corporate organizational level and would only benefit large, proprietary laboratories. For small, nonproprietary laboratories, a cost-accounting system similar to that described in chapter 8, in which deviations from projected expenditures are periodically reported to the laboratory director or manager, should prove useful as a fiscal control technique.

SUBJECTIVE CONTROL TECHNIQUES

Research managers, in many instances, have some level of technical background (Walters 1965). Individuals who have been successful at the "bench" often find that the top of the R & D career ladder leads to a management post. This may be a mixed blessing for those who have little knowledge of management concepts and possess few formal managerial skills. Their technical background, however, provides the basis for one of the best qualitative control techniques—personal observation. Although personal observation is an incomplete control tool by itself, the manager with technical background knows his way around the laboratory and can use this knowledge to gain insight into the progress of a project that might escape the untrained eye. Personal observation provides a strong supplement to other technical and fiscal control techniques.

Drawing from surveys of a number of major companies involved in R & D operations, Dean (1968) has presented data showing that nearly one-half of the respondents reviewed their research projects on a monthly basis. As might be expected, each company placed different emphasis on the essential control variables—that is, technical performance, costs, and progress (timing). A typical control scenario might proceed as follows: Letter progress reports were submitted monthly by group leaders to the laboratory manager; bimonthly informatives (short, informal written reports or memos) were exchanged in large laboratories followed by quarterly meetings in which progress, problems, and fiscal deviations and status were discussed. Accounting reports with expenditure details were usually submitted monthly to higher management followed by quarterly reports of progress and problems. For proprietary laboratories, quarterly reports were normally coordinated with reviews of marketing status. Dean's study illustrates some of the general report formats employed.

TECHNOLOGY TRANSFER AND PROJECT ABANDONMENT

Because the development of most major technologies has consistently required at least eight to twelve years, proprietary companies have developed survival strategies in an effort to maintain their position in a particular marketplace. These strategies usually entail shortening product life cycles and decreasing the intervals between the creation of new products and/or services. To fulfill these goals, technological planning must represent a conscious effort on a company's part to select the best course of action from many possible alternatives. In short, a company must address the need to anticipate or forecast probable directions of technological development.

While we have persistently emphasized the relationship between planning and control techniques, technology transfer represents an area of R & D management in which the two processes merge. It is the planning process that provides the constraints within which the choice of techniques or tools are selected and consequently sets the standards for control of technology transfer or project abandonment.

General categories of technological forecasting techniques include *exploratory* and *normative* forecasting (Fusfeld and Spital 1980). The former is based upon the assumption that a technology progresses in such a manner that, allowing for the impact of appropriate environmental parameters, future positions can be predicted. One method, termed *trend extrapolation,* involves specialized techniques of analysis including trend correlation with precursors, envelop and substitution curves, and technological progress and step curves.

Normative forecasting begins with identification of a particular technological objective and, through such techniques as *relevance trees,* works back to the present. With the relevance tree, objectives are broken down into a hierarchy of tasks, approaches, tools, systems, subsystems, and other elements. By assessing the importance of each task and the importance of each approach to each task, and so on, obstacles are identified and appropriate technology for progress is emphasized. Perspective trees are linked via areas of impact such as environment, technology, valued system attributes, and attributes that may be changed through technological progress. Thus, each tree attempts to define avenues of opportunity and/or sources of hazard.

An intuitive method of technological forecasting that has gained recent favor is the *Delphi* technique. This technique usually involves several rounds of questioning of each of a panel of experts. The first round is designed to elicit predictions of important events that impact upon the area of interest. The experts' responses may simply be brief statements addressed to the question of important developments they perceive as possible or likely to occur in a given future time frame. Statements of all participating experts may then be exchanged and a second round allowed for modification of original predictions and estimates. From these "brainstorming" sessions a general consensus is reached regarding the most significant technological events that will affect a company, their desirability, the likelihood of their development, and the time period in which they can be expected to occur.

In all of these techniques, one immediately recognizes that standards of progress are delineated in a manner by which a controlled transfer of technology can be brought about. The milestones are identified through which the orderly transfer of information must proceed prior to achieving the desired goals.

When managing R & D, one soon realizes that progress does not necessarily proceed according to a set plan—the essence of Murphy's Law. If indeed it did, there would be no need for control measures.

Accordingly, at times some projects must be abandoned. As Dean (1968) observes, "one must be able to recognize when to stop trying to make a good idea work." Some of the previously discussed control techniques are useful when a project abandonment decision must be made. Network switching, with its decision boxes, identifies critical points in the progress of a project and introduces a logical basis for abandonment decisions. In addition, PERT/COST provides the fiscal control for making such decisions. The latter technique plays a major role in abandonment decision making under proprietary circumstances.

SUMMARY

Whereas little can be done to instill innovative and creative attributes, environmental factors do affect the expression of these qualities in individuals who possess them. Indeed, the management of high-talent employees presents some unique problems that we have only recently begun to address. No doubt scientists, engineers, and technicians chaff at the implications that the word "control" conjures and there are those who would argue that the creative individual must be given a free rein if creative talents are to be fully expressed. Others suggest that society can no longer afford the luxury of squandering resources in support of experiments seemingly isolated from reality or outside the established scientific paradigm of knowledge. If the latter attitude prevailed, current scientific dogma would go unchallenged, there would be no occasional "quantum jump" in our knowledge base, the current paradigm would cease to expand, and there would be no future benefit derived from science to society. Nevertheless, we recognize that constraints exerted by resource limitations dictate that society benefit maximally from its expenditures. The integration of innovative and creative attributes into the scientific method provides the framework within which these attributes can be directed toward specific goals. The scientific method of problem solving itself necessitates a high degree of mental discipline. Control, as it relates to R & D, simply represents that managerial function that allows innovation and creativity to be channeled toward specific goals with maximum efficiency. It is encumbent upon the research manager to foster an environment that achieves this end.

Control itself is a process of problem solving. In its basic elements it entails (1) the delineation of objectives, (2) the comparison of performance against those objectives, and (3) correction of any variation from the plan of action that may have occurred. The control variables for research are generally technological performance, costs, and progress (timing). Technological performance and progress are commonly evaluated through specialized techniques of time-event analysis that relate activities to performance times. For less complex projects, Gantt charts and milestone analysis are frequently used. Some type of network analysis involving linear programming techniques, such as PERT, is often employed to control the progress of complex projects where a number of interdependent events are involved.

A research undertaking by its nature carries a high risk of failure. As a consequence, cost considerations have necessitated the application of conventional financial control measures to research efforts. The organizational level at which fiscal control is exerted usually determines the methods employed to control costs. Techniques such as ROI are typically employed at the corporate level, BEP at the divisional level, and PERT/COST at the laboratory level. For small, nonproprietary laboratories, the standards for fiscal control may simply be to maintain an adequate level of funding for the status quo. In any type of setting, financial constraints on a research undertaking make costs an important control variable.

The size of the laboratory, the nature of its research, and its proprietary or nonprofit status will determine the manner by which control is effected. With due consideration of these determinants, the selection of specialized control techniques should be governed by whether they offer expediency, timeliness, flexibility, and economy.

REFERENCES

Braunstein, Y. M., W. J. Baumol, and E. Mansfield. 1980. "The Economics of R & D." In *Studies in the Management Sciences,* ed. B. V. Dean and J. D. Goldhar, vol. 15. 19-32. Amsterdam: North-Holland.

Dean, B. V. 1968. *Evaluating, Selecting, and Controlling R & D Projects.* American Management Study no. 89. New York: American Management Association.

Dean, B. V., and J. D. Goldhar, eds. 1980. "Introduction: Management of Research and Innovation." In *Studies in the Management Sciences,* vol. 15, 1-17. Amsterdam: North-Holland.

Francis, P. H. 1977. *Principles of R & D Management.* New York: AMACOM (a division of the American Management Association).

Fusfeld, A. R., and F. C. Spital. 1980. "Technology Forecasting and Planning in the Corporate Environment: Survey and Comment." In *Studies in the Management Sciences,* ed. B. V. Dean and J. D. Goldhar, vol. 15, 151-62. Amsterdam: North-Holland.

Hodgetts, R. M. 1979. *Management: Theory, Process and Practice.* Philadelphia: Saunders.

Lin, W. T., and M. A. Vasarhelyi. 1980. "Accounting and Financial Control of R & D Expenditures." In *Studies in the Management Sciences,* ed. B. V. Dean and J. D. Goldhar, vol. 15, 199-214. Amsterdam: North-Holland.

McLoughlin, W. G. 1970. *Fundamentals of Research Management.* New York: American Management Association.

Walters, J. E. 1965. *Research Management: Principles and Practice.* Washington, D.C.: Spartan Books.

PART 2
SPECIALIZED ASPECTS OF LABORATORY MANAGEMENT

Chapter 4
Laboratory Safety

Safety. Few words can conjure up a greater number of definitions if one surveys laboratory and supervisory personnel in academic, industrial, or government research laboratories. It brings forward visions of: extra work, a nuisance, a duty to employees, a lot of trouble, government agencies, a proliferation of oppressive regulations, and for every hundred people interviewed—a hundred different responses. It does not have to be this way, however, and it is the laboratory manager who can have the greatest impact on how the people in his laboratory perceive safety.

Space does not permit a complete compendium of safety information. Rather, this chapter is intended to provide the laboratory manager with enough information and ideas to develop his own safety program. At the end of this chapter, the reader will find numerous references to aid him in gathering further information. The references do not form a comprehensive bibliography on laboratory safety, but should be sufficient to cover all areas of concern.

MANAGEMENT OF LABORATORY SAFETY

The people who work in a laboratory tend to assume that their working environment is safe and that they are protected from known hazards. Contrary to these assumptions, the laboratory worker is exposed to a vast array of environmental agents capable of causing acute chronic illnesses and physical injury, as well as the traditional industrial injuries. It is the responsibility of the laboratory manager to bridge the gap between the assumptions and reality, assuring the laboratory worker of the safest possible working environment. The manager's job is to provide safety leadership and instill appropriate attitudes towards safety in the workers (Irving 1971).

Responsibility

The laboratory manager and the supervisory personnel in his laboratory have primary responsibility to ensure the safety of laboratory personnel by preventing accidents and providing safe working conditions. Safety responsibility is often divided into three categories—individual, supervisory, and organizational (Steere 1967).

Individual responsibility in the laboratory includes using learned skills to determine the hazards of an assignment, understanding safe laboratory practices, designing the experiment using proper equipment based on potential hazards, and reporting hazardous conditions or accidents as they are identified. The responsibility of the laboratory manager is to ensure that (1) proper training is provided to each individual in the laboratory; (2) the research proposal is accurate, thorough, and has been evaluated with the health and safety of laboratory personnel in mind; (3) the experimental setup and procedures are properly followed; and (4) proper facilities have been provided. In other words, laboratory safety in relation to the individual can be considered the responsibility of supervisory personnel.

The organizational safety responsibility centers around providing a safe working environment, accident prevention, occupational health programs, and maintenance so that facilities and equipment function properly. The laboratory manager should look on himself as the representative of the organization and its management, the individual responsible for providing safe facilities and properly functioning equipment, as well as the first line supervisor for accident prevention and occupational health programs. Although his may not be the ultimate legal responsibility, this concept helps the laboratory manager understand why it is important for him to be concerned for safety at this level (Schmitz and Davis 1967). He is the one who must report to the organization when there are problems with the facilities or when equipment must be replaced and make sure that all equipment and facilities are properly maintained.

Attitude and Leadership

A laboratory manager will have his greatest impact on safety—both laboratory conditions and personnel actions—by the extension of his attitude and leadership throughout his organization. In moving across the country and studying laboratory safety, safe conditions, and the safety habits of laboratory personnel, several important facts stand out.

- The attitude of management towards safety directly translates to the first line supervision in the laboratory and on to all laboratory personnel.
- A significant asset to a safety program is the strong support of the organization's leadership—chief executive officer, university president, or agency head.
- If management perceives safety as a positive and an inherent quality to good laboratory practices, most personnel in that laboratory will have a like attitude and perception.
- Laboratory managers who require that safety be a part of all research proposals, operational documentation, and experimental descriptions will find that personnel will begin to give safety an automatic (but not thoughtless) role in the development of proposals and procedures.
- As safety topics are found in more facets of their jobs, safety and health considerations will find their way into daily discussions and staff meetings.

In other words, if laboratory management will perceive safety in the laboratory as both positive and an almost subconscious and normal part of all laboratory activities, then this attitude will permeate the organization. Safety and a safe attitude become integrated into the normal routine to the point that it no longer takes special consideration or extra time. A laboratory becomes more productive because there are fewer accidents and the attitude of the individuals towards their jobs and work environment improves dramatically. This is shown in national accident statistics, which can be used to reflect laboratory performance. It can be demonstrated that the number and severity of accidents are lowest in laboratories where management takes a positive stance in discharging their responsibilities towards their employees.

Obviously, it is much easier to say everyone will have a positive attitude toward safety and have a good safety program than it is to implement such a program and keep that attitude alive. At first it takes a great commitment and a lot of effort, and it is not always easy. A safety program is not created by standing up in front of an assemblage of laboratory workers and saying: "Safety is important, and we will all have a positive attitude towards it." There will be problems along the way, and the laboratory manager will have to use a lot of common sense and his best managerial skills to overcome the problems.

No matter how positive management's attitude, there may be an individual who feels that "no one can tell *me*." Closer examination generally shows that this person has the same attitude toward the other rules and regulations of the company, institution, or agency— this is an employee problem that hundreds of books and countless articles have discussed and suggested ways to overcome. Practically speaking, the employee must understand that following the rules is essential to his job and ultimately his employment. The laboratory manager's best option is to follow standard disciplinary procedures for his organization. Whatever actions are taken, the employee must understand that management has resolved to carry out and enforce all of the rules. If one employee is allowed to escape or get away without obeying the laboratory rules, the entire safety program is jeopardized.

Safety Program

One way in which the laboratory manager can clearly carry out his responsibility and demonstrate his concern for laboratory safety is to have an active safety program in his organization. There is no standard for a safety program, but all programs must be flexible and geared around the organization and its personnel. However, certain features are common to most successful safety programs. Some of the most important of these are:

- Published safety procedures for the laboratory
- Established safety programs for all critical health concerns (eyes, face, feet, respiration, and so on)
- An active safety committee
- A program to investigate accidents
- Public dissemination of accident findings to promote accident prevention
- A safety inspection program

An excellent way to promote a safety program is through an orientation program for new employees, including those recently transferred into the department (Irving 1971). Defining the information that is required for an orientation booklet is a good approach to compiling the information needed for the various hazard programs. Such a manual would include the following items:

- Facility diagrams indicating emergency exits and fire alarm locations

- Procedures to be followed in an emergency (fire, tornado, hurricane, and so on)
- Emergency telephone numbers
- Instructions on obtaining medical assistance
- Procedures for reporting an accident
- Instructions on filling out an accident report form
- Standard safety procedures
- Approved practices
- Procedures for obtaining safety equipment

Keep in mind that if the programs and information are laid out in such a manner that the new employee can understand what he is supposed to do, then the information is sufficiently detailed for all employees. One will find that the veteran employee feels less need to review the information and can sometimes have a poor understanding of current procedures. Periodic reviews of established policies should therefore be held for all employees.

An active interest in maintaining safety awareness does not appear at the bequest of the laboratory manager. Safety does not come without effort, and attitude is not sufficient to get the job done. Some employees develop a positive attitude towards safety but at the same time conclude that they do not need to do anything special. They rely on the laboratory manager, their supervisor, the building engineer, or the laboratory administrator to keep their work environment and the laboratory practices safe. This passive attitude will not make safety an integral part of their daily activities. The employee must realize that it takes the active participation of everyone to keep the workplace safe.

A leading cause of a passive attitude is lack of experience in taking responsibility. The best way to correct that is to break responsibilities into smaller tasks and assign them to different individuals. Rotate these assignments so that every member of the laboratory at one time or another during a year has some active responsibility in ensuring a safe working environment for all the employees. The tasks do not have to be large or time consuming, and the employee will develop an appreciation for what it takes to maintain a safe laboratory. With time, the laboratory manager will find some employees quite adept at carrying out their assigned tasks, and he should continue to draw on these individuals in order to help maintain the momentum of the safety program. There will be other individuals who will have difficulty carrying out safety tasks, either because of a lack of skills or organizational ability, or because of indifference or negativity. These individuals will need as much guidance as possible and it will be best to give them smaller and simpler tasks. However, no one in the laboratory should be allowed to escape having some responsibility for ensuring laboratory safety—if for no other reason than to develop an appreciation of how important the management of safety is.

Safety Committee

Forming a safety committee is an effective way of gaining participation by laboratory members and disseminating safety information throughout the laboratory (Loperfido 1974). While there are committees at the company, university, or hospital level, the department-level committee plays an essential role in promoting safety where it is most important. There are probably as many ways to organize such a committee as there are committees, and one method is not necessarily better than any other. What the laboratory manager must do is look for the type of organization that will be effective in carrying out its intended duties in his laboratory. If the committee is not functioning effectively, then the proper course is to reorganize it. This is necessary because an ineffective safety committee will continue to become less active until it ceases to function. Therefore, look for a structure that calls for participation by individuals who are interested and anxious to accept the duties. For a given laboratory this could mean representation by supervisors, technicians, or members of the senior research staff. Representation can be made up from each functional area or by academic disciplines. However representation is formed, make sure that it is working.

An active safety committee can take a major burden off the laboratory manager by assuming the following types of responsibilities:

- Conducting safety inspections of the laboratories
- Investigating accidents
- Disseminating current safety information that is important to members of the laboratory staff
- Reviewing safety procedures and policies
- Conducting training programs for employees on potentially hazardous equipment or procedures
- Arranging for films and slide presentations for laboratory personnel
- Conducting orientation programs for new employees

A laboratory manager will find that the more responsibility is given to the safety committee, the more duties it will assume, thus aiding the entire organization.

that are required. When some action has been taken, let the employee that reported the accident or near miss know that something has been done. The quickest way to end conscientious reporting of accidents is to take no action or delay taking action. Write a formal report on each investigation. If your organization has no form for reporting accidents, you can consult any of the available references (such as Steere 1971*a*, Fawcett 1965, and National Safety Council 1965) for help in designing one for your own needs. Although forms differ, they all provide basically the same information:

- Who had the accident?
- When did it happen?
- Where did it happen?
- What actions were taken by the person involved?
- What inflicted the injury or damage?
- What happened?
- What things caused the accident?
- How can this type of accident be prevented in the future?

It is important to isolate the true causes of the accident and either correct the problem condition or use the accident to inform laboratory personnel about actions that could be harmful.

Safety Audits and Inspections

As a part of the occupational health program in all organizations with laboratories, an annual safety audit is performed by experienced and knowledgeable personnel. This audit serves both the safety and legal requirements of formally reporting any unsafe conditions found in laboratory areas, specifying steps to be taken to correct these conditions, and following up on conditions identified in previous audits (Wood 1974). The laboratory is a dynamic place, and an audit once a year with brief reviews of identified problems cannot possibly uncover all unsafe conditions. The safety inspection is therefore an excellent way to complement the annual audit. It provides an additional active role for laboratory personnel, supplementing the safety program and the work of the safety committee. An inspection team should carry out periodic examinations of the laboratory areas, offices, specialty rooms, and storage areas to check for possible safety-related problems and hazards. The inspection team should be made up of laboratory personnel and supplemented with individuals with a strong knowledge of safety codes and industrial hygiene. These knowledgeable people could be employees of another department or individuals recommended by the local fire marshal. It is not a difficult task to develop a safety inspection form, and many chemical companies and safety supply companies have such forms available. These forms can be easily customized for each laboratory. New members of the inspection team should be taught what to look for, and it is a good idea constantly to groom new individuals for the inspection task.

Internal safety inspections should be carried out periodically at intervals determined by the conditions in the laboratory. When instituting a new inspection program, it might be necessary to inspect all laboratories once a month. However, where there is an active safety program, quarterly inspections may be sufficient. Regardless of the committee makeup or the frequency of the inspections, they will uncover potential hazards that need correction. The committee should also be charged with following up on cited problems in order to ensure that the conditions are corrected. The safety inspection becomes an excellent tool to keep interest high and maintain participation by a number of individuals. Whether the inspection is quarterly or even monthly, it will not uncover all hazards that can take place in a dynamic facility. Conditions, procedures, and experimental setups change. The personnel in the laboratory must not become complacent because there is an inspection program. Members of the inspection team and the supervisory personnel should be constantly watching for potential hazards and be cognizant of code violations at all times. Everyone must understand that the time to cite a problem is when it is found, not during the next safety inspection.

LABORATORY DESIGN AND EQUIPMENT

Providing a safe working environment for a laboratory, with its inherent and produced hazards, is not easy. Careful planning is needed to ensure that it is properly designed, meets all code requirements, and includes the necessary emergency and protective equipment. There is no magic list that defines the facilities and equipment required for a given laboratory. The requirements will have to be developed based on the materials to be used and the experiments that are to be carried out. Fire prevention codes and local ordinances require major items, such as building sprinkler systems, fire alarms, building emergency alarm systems, fire extinguishers, emergency exits, safety showers, and eye wash outlets. To ensure that adequate protection is provided, the laboratory manager should consult with the facility safety manager, the facility's insurance carrier, the local fire marshal, or a professional engineer. Other commonly required equipment includes fire blankets, stretchers, gas masks, and respirators. Safety equipment can be divided into two broad categories—emergency and protective.

Laboratory Design

Laboratory design can work with and strengthen the safety program that the laboratory manager has worked hard to implement, or it can have a negative impact, destroying momentum and interest. An improperly designed laboratory can indeed be a hindrance—if material flow requires individuals to walk around each other frequently, if the location of central services requires half the laboratory to walk by one individual's bench, or if lighting or ventilation is inadequate for the experimental work. The laboratory that is well laid out—designed with material flows and experimental procedures in mind—can enhance all aspects of a safety program. Unfortunately, laboratories are not redesigned every time a new program starts, and rarely does the laboratory manager have the opportunity to start from scratch and design a facility based on his group's needs. It is necessary, therefore, to work within the constraints of the facilities that are provided. Working within this limitation, the laboratory manager should correct major problems as quickly as possible. If a major renovation is called for, the manager will need to determine exactly what changes should be made, ranking the tasks in priority according to safety, hazard, and increased functionality. This information can then be used to set a long-term goal for general laboratory renovation. Whenever maintenance is performed, new equipment is installed, or old equipment is removed, you should examine the action in terms of the long-range plans for the laboratory and carry out maintenance that fits in with these goals. Whenever funding is appropriated for minor renovation projects, use the long-term laboratory plan as an aid in determining which areas should be renovated first. If appropriations are made for a major renovation, this plan will aid the laboratory manager in discussing the needed changes with architects and engineers.

Sometimes the laboratory manager has little influence on design because he inherits a very old or overcrowded laboratory. Legal and moral responsibilities do not allow this to be used as an excuse. Unless there are physical deficiencies such as inadequate ventilation, code violations, or the unsafe installation of equipment, there is always something that the laboratory manager can do to improve conditions. When there are code violations or the ventilation is inadequate, there is a responsibility to raise a red flag to alert the institutional management, so they can understand what is needed to correct the problems. The laboratory manager who does not raise these issues is evading his responsibility to both his employees and his employers.

Most problems can be corrected with a little work. An engineer can help the laboratory manager reroute utilities, increase electrical capacity, rearrange equipment so that no egress is blocked, provide adequate sprinklers and fire extinguishers based on the activities and hazards present in the laboratory, and in general, ensure that all code requirements—local, state, federal, National Fire Protection Association (NFPA), and insurance carrier—are followed. The manager will need to discuss problems and conditions with an engineer. If the laboratory manager can find an engineer with whom he can develop a rapport, it becomes a very fruitful experience. When a manager takes charge of a laboratory that has evolved over many years, correcting safety problems can look like a monumental task, and it sometimes proves to be just that. If a good engineer is consulted regularly during the process of rearranging and reallocating space, it will prove less difficult. There is no guarantee, however, that the work will be covered by the existing budget.

This is especially true in the older laboratory that has evolved to its present state without major renovations. In this type of setting, changes compound problems, which compound code violations, leading to a mass of difficulties that appears insurmountable. The good laboratory manager cannot throw up his hands and say "impossible!" Using available resources and his budgeted funds, he must attack each problem, beginning with the one that has the greatest potential for hazard. Under these circumstances, it is helpful to be honest with the employees. Describe the problem conditions, what management is doing to correct them, and the anticipated date for completing the changes. Review with the employees all preventive measures that can be taken in the meantime to minimize danger. Provide special wearing apparel, such as rubberized aprons and protective boots and additional safety equipment, such as goggles and shields. The innovative laboratory manager can find ways to create a safer working environment even where problems exist.

Some laboratories just look unsafe. Inspections do not find any code violations in engineering, but one constantly finds blocked aisles, spills, and unused equipment lying around. This describes the root of the most commonly cited violation in laboratory inspections— the lack of good housekeeping. Good housekeeping must be practiced continually to prevent accidents, fires, and injuries. Individuals in the laboratory must be expected to keep their work area clean, neat, and orderly. The practice of cleaning up after each phase of an experiment and immediately after any spill will lessen the need for time-consuming major cleanups. These are some rules to follow:

- Put away any materials that are not currently being used.
- Clean up benches, tables, and hoods after each experimental step.
- Do not pile boxes or leave papers strewn on floors or in aisles.
- Keep the passageway to all exits clear.
- Mark off an area around safety showers, fire extinguishers, fire blankets, and eye washes, and do not allow anything to block these areas.
- Clean up all spills immediately and use approved methods to dispose of materials that were used to absorb the spills.
- Wipe up or pick up any tripping hazards you find on the floor, such as dropped ice, stir rods, pencils or erasers.
- Dispose of all broken glassware in a clearly marked container.
- Segregate your waste receptacles for solvents, glassware, radioactive materials, paper, and other materials.
- Keep the outside of all chemical containers clean and ensure that the label is correct and easily read.
- Keep in your work area only the chemicals needed for work in progress—do not create your own chemical storage area.
- Have all equipment not in current use stored so that it does not clutter your work area.
- Do not pile boxes or other materials to dangerous heights.
- Do not store jackets, sweaters, lab coats, or other garments on benches or hang them on cylinders—put them in your locker, closet, or other appropriate storage area.

When in doubt about housekeeping in the laboratory, consider the old adage—"a place for everything and everything in its place."

Emergency Equipment

Emergency equipment is designed to protect life, limb, or property, and the use of it requires quick access and prompt action (Steere 1971). Each person working in the laboratory should know where each piece of emergency equipment is located and how to use it. Training should be provided to all laboratory personnel on the actual use of most pieces of equipment. Periodic refresher courses prove very beneficial. Do not assume that because an employee has transferred from another department or laboratory he knows what equipment exists, where it is located, and how to use it. The transferred employee should go through the same training as any new employee. Once a year, hold an emergency training course for all laboratory personnel. At the very least, have hands-on drills using fire extinguishers. Local fire marshals are generally more than willing to provide training, if no one inside your organization has the expertise to conduct the fire emergency training program. The training should include, at a minimum, fire evacuation drills, the use of fire blankets, and some rudimentary levels of first aid—including the use of a stretcher. Most first aid activities, however, should be left to a trained squad inside the organization or to local paramedics. The use of spill cleanup kits for chemical spills should be part of the training program during new employee orientation. An annual refresher helps as the individual becomes familiar with additional procedures and uses new chemicals. Gas masks may be ineffective, due to their short life, as well as potentially fatal because they lead to a false sense of security on the part of the user. The only type of respirator that should be made available is a self-contained breathing apparatus (SCBA). The training and use of this equipment should be left to a select few individuals, probably members of a rescue squad, as it requires frequent refresher courses and physical agility. An additional hazard is created when training has not been offered on the use of safety equipment that is available—the laboratory staff may develop a false sense of security.

Special Equipment and Facilities

The other major category of safety equipment is personal protective equipment. This includes articles such as safety glasses, goggles, face shields, gloves (rubber for handling corrosive chemicals or solvent-resistant plastic for handling solvents), rubber aprons, and proper footwear. Canvas or open-toed shoes are never acceptable. Each individual should wear substantial shoes, and when laboratory conditions require, steel-toed safety shoes should be supplied. Other types of personal protective equipment are provided as part of the facility. Among these are shatter-proof shields, fume hoods, bottle carriers, biohazard hoods, radiation hoods, and glove boxes. Safety glasses should be required wearing apparel for everyone (including secretaries and visitors) in any laboratory where there is a potential for the splashing or spilling of a chemical. The development of an eye protection program defining the use of goggles and shields for certain procedures is a positive approach to ensuring that each employee is protected against eye and face injury. The wearing of contact lenses is generally discouraged, but if they are worn, goggles or a full face shield must be used.

Fume hoods are rated according to the activity and type of hazard that can be contained (NFPA 1975). Table 4-1 can be used as a guide to determining the quality of hood required, according to the nature of materials handled. Note that the older hoods that are still in use cannot handle the higher air velocities. Some of the newer hood designs, with higher capture efficiency, can use lower face velocities. Consult the manufacturers or a qualified engineer for help in defining your requirements. It is important that each hood be clearly labeled so that the employee can determine whether his experiment is appropriate for the hood and not develop a false sense of security. Biohoods and biocontrol stations are also designed according to the type and classification of hazard that might be encountered, as shown in table 4-2 (Richardson and Barkley 1983). For these devices to work properly, the employee must understand his experiment and the nature of the potential hazard. More importantly, he must understand the design criteria and limitations of the safety devices. Only in understanding both areas can you be assured that the employee will use proper judgment. The employee should be instructed to ask his supervisor if there is any doubt.

HAZARDOUS MATERIALS

The routine use of hazardous and potentially hazardous materials in the laboratory presents an ongoing problem for the laboratory

Table 4-1. Laboratory Fume Hood Face Velocities

Nature of Materials Handled	Industrial Hygiene Threshold Limit Value Standards			Actual Face Velocities	
	Gases, Vapors (ppm)	Dusts, Fumes, Mists (mg/m^3)	Mineral Dusts (mppcf)	Avg. (fpm)	Min. (fpm)
Extremely toxic	<10	<1	<20	125-150	110-125
Very toxic	10-100	1-5	20-30	100-125	80-100
Slightly toxic	100-1000	5-10	30-40	80-100	60-80

SOURCE: National Fire Protection Association, 1975.

76 SPECIALIZED ASPECTS OF LABORATORY MANAGEMENT

Table 4-2. Safety Performance Requirements and
Specifications of Biological Safety Cabinets

Cabinet	Recombinant DNA Facility Classification	Face Velocity (linear ft/min)	Leak Tightness
Class I	P1–P3	75	Not Applicable
Class II, Type 1	P1–P3	75	Gas tight; Leak rate < 1×10^{-5} cc/sec @ 2″wg pressure
Class II Type 2	P1–P3	100	Pressure tight; No air/soap bubble @ 2″wg pressure
Class III	P4	Not Applicable	Gas tight; Leak rate < 1×10^{-6} cc/sec @ 3″wg pressure

Performance Requirements[a]

SOURCE: Richardson and Barkley, 1983.
[a] All performance requirements are based on using HEPA filters with 99.97 percent efficiency.

manager. He cannot be present or oversee all of the ongoing activities, so he must rely on training, the development of a safety awareness by his workers, and the proper review of safety in research proposals and operational procedures. If the laboratory facilities are well designed, he can then be reasonably assured that he has done what is in his power to provide for a safe working environment. A remaining responsibility is to ensure that safe operating procedures are established and followed for handling hazardous materials. While more than adequate information exists on most chemical hazards, the laboratory manager must be concerned with all the newly identified hazards and the ongoing changes in federal and state regulations. For that reason, it is best to have a trained professional in the organization who is dedicated to monitoring these changes and alerting the entire staff. Where this must be considered a luxury, it would be adequate for a laboratory manager to appoint several individuals to monitor activities in their own specializations. When in doubt, the best policy would be to follow

1. Current federal regulations by departments such as National Institutes of Health, Center for Disease Control, Environmental Protection Agency, Department of Transportation, Department of Labor, Occupational Safety and Health Administration, and National Institute for Occupational Safety and Health

Table 4-3. Organization of Chemical Storage in Limited Spaces

Inorganic Chemicals	Organic Chemicals
Sulfur, phosphorus, arsenic, phosphorus pentoxide	Alcohols, glycols, amines, amides, imines, inides
Halides, sulfates, sulfites, thiosulfates, phosphates, halogens	Hydrocarbons, esters, aldehydes
Amides, nitrates (except ammonium nitrate), nitrites, azides, nitric acid	Ethers, ketones, ketenes, halogenated hydrocarbons, ethylene oxide
Metals, hydrides	Epoxy compounds, isocyanates
Hydroxides, oxides, silicates, carbonates, carbon	Sulfides, polysulfides, sulfoxides, nitriles
Arsenates, cyanides, cyanales	Phenols, cresols
Sulfides, selenides, phosphides, carbides, nitrides	Peroxides, hydroperoxides, azides
Borates, chromates, manganates, permanganates	Acids, anhydrides, peracids
Chlorates, perchlorates, perchloric acid, chlorites, hypochlorites, peroxides, hydrogen peroxide	Miscellaneous
Acids (except nitric)	Miscellaneous (nitric acid)

2. Established practices by such organizations as Chemical Manufacturers Association and American Chemical Society
3. Guidelines and safety information provided by chemical manufacturing companies such as J. T. Baker, Fisher Scientific, and American Scientific Products

Storage

Storage of chemicals is an ongoing problem. The goal is to find a cost-effective, efficient, space-conserving, and safe way to store them. It would be ideal if each class of chemical could be stored in a separate facility, but most of the time this is not practical. Arranging by alphabetical order is probably the worst method of organizing a storage area. Although several sophisticated methods of organization exist, many laboratories manage with the minimum requirement of only two storage cabinets. This assumes that flammable liquids are stored in an approved solvent storage cabinet and toxicants are isolated from other chemicals in a cool, dry, well-ventilated area away from heat, moisture, and sunlight. Table 4-3 shows just one possible storage pattern for laboratories or small storage rooms where space is

limited. Since such confined storage is not ideal, it is best to plan adequate storage during laboratory design. Although most laboratories do not have all the categories of chemicals shown, every laboratory uses inert chemicals, which can best be stored on the upper shelves. Two rules are essential:
1. DO NOT store corrosives above shoulder level.
2. DO NOT store chemicals on the floor.

A major concern for the laboratory manager is not to let storage get out of hand. Question the researchers concerning the volume of orders. Do they really need that much on hand or can they arrange drop shipments periodically with the vendor and avoid large storage requirements? Periodically review the storage areas, asking these questions:

- Are there old chemicals present?
- Are they still needed?
- Have they lost their effectiveness?
- Is there a potential hazard present because of reactions in aged chemicals (for example, peroxide-forming chemicals such as ethers)?

A chemical inventory system should be considered mandatory, whether it is kept manually or automated on a computer. The chemical inventory should indicate quantities, date received, storage location, chemical grade, and vendor. Keeping an up-to-date inventory is an excellent way to control the accumulation of chemicals. It will also prove useful in responding to periodic regulatory questions concerning toxic and hazardous chemicals. The inventory should always be checked prior to ordering additional hazardous materials.

Handling

When handling materials, the worker must be concerned not only with the obvious warnings that apply to toxic and corrosive chemicals but also with the potential hazards of numerous inert agents. For toxic and corrosive chemicals, there are a few simple rules:
1. Handle carefully.
2. Wear adequate protective equipment when using.
3. Use fume hood or other exhaust system that is available.
4. Wash thoroughly after use.
5. For leaky containers, take extra precaution, wear additional protective equipment, isolate the leak, and report it immediately.

Since the hazard associated with many inert substances is not as clear, additional precautions are often needed. One type of hazard results from the many common laboratory practices that cause dusting and aerosoling. Fine airborne particles frequently come to rest on scattered surfaces around the laboratory and on workers. Landing on bare skin, they can cause delayed reactions—dermatological conditions ranging from mild rashes to serious lesions. More serious is the potential for reaction with sensitive tissues such as the mouth, esophagus, nose, mucal membranes, eyes, and bronchial passageways. Workers may experience severe respiratory attack, sometimes requiring hospitalization. Proteins present in airborne particles can cause some of these reactions. The responses to some chemicals are less well-documented and therefore more difficult for the laboratory manager to anticipate. The lack of clearly established guidelines and classification makes these reactions unpredictable.

The following rules will help to control the hazards posed by handling chemicals:

1. When new activities are started in a laboratory, make sure there has been a thorough safety review of all the components and tasks.
2. Make sure that all the safety procedures are written and followed for tasks that might have a potential hazard.
3. Provide safety equipment as needed, such as laminar work stations, enclosed weigh stations, and special ventilation.
4. Provide employees with safety apparel, such as laboratory coats, gloves, respiratory masks, and goggles.
5. Have procedures established on what to do in the event of an emergency, just as you would have for a toxic chemical spill.
6. If the hazards of a particular material are not known, treat it as if it is very hazardous.

Transportation

It is the responsibility of the laboratory manager to ensure that chemicals that must be moved from one location to another are packaged and transported in accordance with Department of Transportation regulations. Materials should be transported in the safest and most expedient manner, i.e., commercial transport in accordance with DOT guidelines. A private automobile should be the last choice. Nonhazardous materials should be packaged in double containers—plastic, if possible. Small containers should be packed tightly in a box, to keep them from moving around. The container should carry instructions on what to do in the event of breakage or spills. For hazardous chemicals, rules and regulations become quite complex, varying by size of container, mode of transportation and classification of hazard. When specific information is required, the reader should consult Title 49 of the Code of Federal Regulations. Part 172.101 contains the hazardous materials table with columns specifying labels, packages, and exceptions.

Disposal of Wastes

The waste disposal problem has grown with the increase in environmental consciousness and the proliferation of regulations at the federal, state, and local level. The laboratory manager can be held responsible for wastes from his laboratory, and the penalties can be severe. It is thus important to follow laws, guidelines, and common sense every time materials are disposed. Laboratory personnel should understand the hazards of chemicals before they are ordered. It is a good idea to anticipate any disposal problems at the project development stage so that the research proposal contains a complete list of raw materials, products, by-products, and chemical wastes along with the hazards of each and methods of neutralizing the waste products. In this way the laboratory manager knows what procedures will be required and he can keep the organizational management informed so they can be prepared for the wastes before they must be disposed. Wastes can be divided into several categories for handling:

- Trash
- Sharps
- Serum
- Biologicals
- Chemicals
- Solvents

Trash includes waste paper and other nonchemical refuse including that collected from office and cafeteria areas. It should be isolated from hazardous wastes, and no special handling precautions are required. Chemicals, glass and other sharp objects, serum, and biologicals should never be disposed of with the trash.

Sharps, including glass, syringes, pipettes, and short metal objects, should be placed in a metal container labeled "Glass Only." This is a precaution for both the researcher and janitorial crews. Statistics show that one of the most common accidents among janitorial service personnel in laboratories is a puncture wound from a sharp object in a trash container.

Waste solvents should never be flushed down a drain. They should be separated and stored for proper disposal. Due to their flammability, they should be disposed of in fire safety containers located at strategic points in the laboratory and marked "Waste Solvents Only." Laboratory personnel can pour their neutralized waste solvents into these cans, thereby keeping to a minimum both the fire potential and the quantity of waste solvents in the laboratory. The waste solvent cans should be periodically emptied into drums kept at a remote site from the laboratory and other occupied buildings. A professional waste firm should be contracted to handle the disposal of bulk waste solvents.

Waste chemicals that are nontoxic and neutralizable can, in most cities, be flushed down the drain with enough water to ensure adequate dilution. Waste chemicals that cannot be flushed down a drain must be disposed of in an EPA-approved landfill. If practical, it would be a good idea to store waste chemicals in a remote site away from occupied buildings until there is a sufficient quantity to justify bringing in a professional waste-handling company.

SPECIAL TOPICS

A good laboratory manager can go to great lengths to provide a safe working environment for his employees, but find that a project will come along that requires additional precautions. All the advanced planning, procedures, and facility design efforts will turn out to be inadequate for some new projects. If any additional requirements are identified during planning and the development of the research proposal, there will be time to prepare new operating procedures and make the necessary facility changes. Two classifications that require such special handling include biohazards (including some microbiological agents, some enzymes, materials determined to be carcinogenic, and teratogenic agents) and radioactive hazards. These will be briefly reviewed in order to aid the laboratory manager in making the appropriate decisions.

Biohazards

In activities where biohazardous materials will be utilized, it is important to provide the necessary containment barriers based on the nature of the agent, the process, the quantity of viable material, and the handling methods. Primary containment provides for the protection of personnel in the local laboratory environment by the use of good microbiological techniques and appropriate safety equipment. Secondary containment provides for the protection of the external environment and the facility outside of the contained area. This is provided by a combination of operational procedures and facility design. Biosafety requires a combination of good laboratory techniques, proper safety equipment, and appropriate facility design in order to reduce the chances of exposure of laboratory personnel, nonlaboratory personnel, and the outside environment to any potentially hazardous agents.

The most important aspect of containment is the adoption and practice of good microbiological and laboratory techniques. Each individual must be aware of the potential hazards and must have been trained in appropriate techniques for the safe handling of the materials. The laboratory manager is ultimately responsible for ensuring that all his personnel have been properly trained. There should be an operational procedures manual that identifies the hazards, biosafety techniques, containment validation procedures, and emergency procedures. Biological safety cabinets and biocontainment stations should be provided whenever good microbiological techniques are not sufficient to protect the laboratory worker. Personal protective equipment such as gloves, gowns, and face shields should be provided if required for the procedures at hand. Always be cognizant of procedures that could cause aerosoling of the harmful agents.

Facility design is essential to providing a barrier to protect other personnel in the facility as well as those in the community from the accidental release of viable agents. The laboratory manager is responsible for determining the adequacy of the facility for the project to be carried out. If the facility is inadequate for the project because of age or design, he must determine alternative procedures or sites for the experimental work. Information on the four biosafety levels that have been identified for infectious agents is provided in table 4-4.

Radiation Hazards

Radioactive materials contain chemicals that spontaneously produce radioactive emissions. The potential for hazard is based on the type of emissions, quantity of material, the length of time of exposure, and the age of the individual receiving the exposure. All radioactive materials should be considered hazardous and treated accordingly. The materials, including the radioactive wastes, should be kept isolated. The use of radioactive materials requires licensing through the Nuclear Regulatory Commission. The laboratory manager should have a radiological control officer who handles all licensing procedures and is knowledgeable about the protective barriers that are necessary to

Table 4-4. Summary of Recommended Biosafety Levels for Infectious Agents

Biosafety Level (Facility Type)	Practices and Techniques	Safety Equipment
1 (Basic)	Standard microbiological practices	None: primary containment provided by adherence to standard laboratory practices during open bench operations
2 (Basic)	Level 1 practices plus: Laboratory coats; decontamination of all infectious wastes; limited access; protective gloves and biohazard warning signs as indicated	Partial containment equipment (i.e. Class I or II biological safety cabinets) used to conduct mechanical and manipulative procedures that have high aerosol potential that may increase the risk of exposure to personnel
3 (Containment)	Level 2 practices plus: special laboratory clothing; controlled access	Partial containment equipment used for all manipulations of infectious materials
4 (Maximum containment)	Level 3 practices plus: entrance through change room where street clothing is removed and laboratory clothing donned; shower on exit; all wastes decontaminated on exit from facility	Maximum containment equipment (i.e. Class III biological safety cabinet or partial containment equipment in combination with full-body, air-supplied, positive-pressure personnel suit) used for all procedures and activities

SOURCE: Richardson and Barkley, 1983.

provide a safe working environment. This individual would work with the waste disposal firm for the proper disposal of radioactive wastes, as well as instructing laboratory workers in decontamination of bench surfaces, glassware, and other utensils; the handling of contaminated skin or lab coats; and cleanup after spills.

Working Alone

Some laboratory personnel want to come into the laboratory at night or on weekends in order to follow experimental results or check on experiments in progress (Reider 1974). As a rule, employees should never be allowed to work alone—in the laboratory by themselves, out of visual or audio contact with another individual, or when no one else is aware of the individual's activities. Obviously, some laboratory work does not stop at the end of the work day or on weekends and must be monitored periodically. So, individuals have a need to gain access to their laboratory during the off hours. Since it is often impractical for another person to be present, some flexibility must be established.

Require any individual working after other than normal work hours to sign a register when entering and leaving the facility. In this way, guards or rescue workers will have some knowledge of the presence of individuals in the facility. Researchers should carry out only routine procedures when working alone. When no one else is present, personnel should not work with flammable or hazardous materials. An employee should not be allowed to work more than a specified length of time without at least telephone contact with another individual. If there is doubt about the safety of the employee, another worker should be assigned to the same shift or the work should be done during normal work hours.

SUMMARY

Safety in the laboratory is not cut and dried. There are as many facets to safety as there are hazards present. It is not an easy task for the laboratory manager, but an essential one. The key elements are to develop a positive attitude among workers; provide a positive leadership role; let all the workers understand his genuine concern for their health and well-being; establish flexible procedures for the laboratory personnel; and encourage an active safety program so that safety consciousness is not a burden, but rather an integral part of the work environment.

REFERENCES

Irving, J. R. 1971. "A Laboratory Safety Orientation Lecture for the First Chemistry Course." *Safety in the Chemical Laboratory* 2:5ff.

Loperfido, J. C. 1974. "Development of a Safety Program for Academic Laboratories." *Safety in the Chemical Laboratory* 3:11ff.

National Fire Protection Association. 1975. *Laboratories Using Chemicals.* No. 45. Boston.

Reider, R. 1974. "Working Alone in Research and Development Activities." *Safety in the Chemical Laboratory* 3:16ff.

Richardson, J. H., and W. E. Barkley, eds. 1983. *Biosafety in Microbiological and Biomedical Laboratories.* Draft. Bethesda, Md.: National Institutes of Health.

Schmitz, T. M., and R. K. Davies. 1967. "Laboratory Accident Liability: Academic and Industrial." *Cleveland-Marshall Law Review* 16:75ff.
Steere, N. V. 1967. "Responsibility for Accident Prevention." *Safety in the Chemical Laboratory* 1:1ff.
_____, ed. 1971. *CRC Handbook of Laboratory Safety.* 2d ed. Boca Raton, Fla.: CRC (Chemical Rubber Company) Press.
_____. 1974. "Identifying Multiple Causes of Laboratory Accidents and Injuries." *Safety in the Chemical Laboratory* 3:1ff.
Wood, W. S. 1974. "A Safety Survey in Research and Laboratories." *Safety in the Chemical Laboratory* 3:28ff.

GENERAL BIBLIOGRAPHY

American Chemical Society. 1974. *Safety in the Academic Chemical Laboratory.* Washington, D.C.: ACS.
Fawcett, H. H., and W. S. Wood. 1965. *Safety and Accident Prevention in Chemical Operations.* New York: John Wiley.
Federal Register. Code of Federal Regulations (CFR) Title 10: Radiation. Washington, D.C.: U.S. Government Printing Office. Updated regularly.
_____. *CFR Title 21: Occupational Safety and Health.* Washington, D.C.: U.S. Government Printing Office. Updated regularly.
_____. *CFR Title 40: Environmental Protection.* Washington, D.C.: U.S. Government Printing Office. Updated regularly.
_____. *CFR Title 42: Public Health and Biosafety.* Washington, D.C.: U.S. Government Printing Office. Updated regularly.
_____. *CFR Title 49: Transportation.* Washington, D.C.: U.S. Government Printing Office. Updated regularly.
Gaston, P. J. 1964. *The Care, Handling, and Disposal of Dangerous Chemicals.* Aberdeen, Scotland: Institute of Science Technology, Northern Publishers.
Guelich, J. 1956. *Chemical Safety Supervision.* New York: Reinhold Publishing.
Lewis, H. F. 1962. "Laboratory Planning for Chemistry and Chemical Engineering." *Occupational Health and Safety,* ch. 7. New York: Reinhold Publishing.
Manufacturing Chemists Association. 1970. *Laboratory Waste Disposal Manual.* 2d ed. Washington, D.C.
_____. 1972. *Guide for Safety in the Chemical Laboratory.* 2d ed. New York: Van Nostrand Reinhold.
Morgan, K. Z., and J. E. Turner, eds. 1967. *Principles of Radiation Protection: A Textbook of Health Physics.* New York: John Wiley.
National Fire Protection Association. 1975a. *Hazardous Chemicals Data, 1975.* No. 49. Boston: NFPA.
_____. 1975b. *Laboratories Using Chemicals, 1975.* No. 45. Boston: NFPA.
National Institutes of Health. 1978. *Laboratory Safety Monograph: A Supplement to the NIH Guidelines for Recombinant DNA Research.* Bethesda, Md.: NIH.
National Research Council. 1981. *Prudent Practices for Handling Hazardous Chemicals in Laboratories.* Washington, D.C.: National Academy Press.

_____. 1983. *Prudent Practices for Disposal of Chemicals from Laboratories.* Washington, D.C.: National Academy Press.
National Safety Council. 1964. *Accident Prevention Manual for Industrial Operations.* 5th ed. Chicago: NSC.
_____. 1976. *R & D Safety Notebook.* Fact Sheets and Data Sheets issued by R & D Section. Chicago: NSC.
Patty, F. A. 1958. *Industrial Hygiene and Toxicology.* 2d ed., vol. 1. New York: Interscience Publishing.
_____. 1962. *Industrial Hygiene and Toxicology.* 2d ed., vol. 2. New York: Interscience Publishing.
Pieters, H. A. J., and J. W. Creyghton. 1957. *Safety in the Chemical Laboratory.* 2d ed. Washington, D.C.: Butterworth.
Powers, P. W. 1976. *How to Dispose of Toxic Substances and Industrial Wastes.* Park Ridge, N.J.: Noyes Data Corp.
Quam, G. N. 1963. *Safety Practice for Chemical Laboratories.* Villanova, Pa.: Villanova Press.
Richardson, J. H., and W. E. Barkley, eds. 1983. *Biosafety in Microbiological and Biomedical Laboratories.* Draft. Bethesda, Md.: National Institutes of Health.
Sax, N. I. 1963. *Dangerous Properties of Industrial Materials.* New York: Reinhold Publishing.
Steere, N. V., ed. 1967. *Safety in the Chemical Laboratory.* Vol. 1. Easton, Pa.: *Journal of Chemical Education,* Division of American Chemical Society.
_____. 1971a. *CRC Handbook of Laboratory Safety.* 2d ed. Boca Raton, Fla.: CRC (Chemical Rubber Company) Press, 1971.
_____. 1971b. *Safety in the Chemical Laboratory.* Vol. 2. Easton, Pa.: *Journal of Chemical Education,* Division of American Chemical Society.
_____. 1974. *Safety in the Chemical Laboratory.* Vol. 3. Easton, Pa.: *Journal of Chemical Education,* Division of American Chemical Society.
Weast, R. C., ed. *Handbook of Chemistry and Physics.* Cleveland: CRC (Chemical Rubber Company) Press. Published annually.

Chapter 5
Effective Technical Communications

Communications are as essential to technology as any of the learned scientific skills. No matter how many entrepreneurs or scientists working alone in a lab have made far-reaching discoveries or provided new products to make life better or easier, they have needed other people. In reality, every innovator has used knowledge that was provided by another individual or group. Furthermore, the value of a discovery is limited by the ability to communicate that information to the outside world so as to make the product useful or the information understandable. It is thus incumbent on every technically trained person to learn the necessary skills to communicate his knowledge, his work, and the results of his efforts to others.

Technical communications come in many forms, both written and verbal: written reports in the form of letters, memoranda, internal documentation, and project reports; articles in technical journals; and articles in lay publications. There are seminars presented before members of professional organizations or closely related fields, presentations to management, and talks given to nontechnical audiences. It is very important for the laboratory manager to understand the needs of the listening or reading audience and the intended use of the information before setting out to communicate with others. Even a very good technical writer and speaker will need to develop different communication techniques to reach a management audience.

The research manager has a special obligation to his superiors, to those working for him, and to the public to ensure that communications originating from his department are well prepared, clear, concise, understandable, and not obscure to the audience for which they are intended. This responsibility might require providing continuing education for employees so that their communications skills are at a level help his employees achieve all deserved recognition. Besides the benefit to the employee, it provides additional recognition for the organization, which can help in such ways as recruiting new personnel or providing funding for future projects.

The laboratory manager should take certain precautions before papers are published, however, in order to protect the legal and business interests of the organization. All papers should be reviewed to ensure that there are no vague areas or underlying meanings that could misinform a peer, confuse the lay reader, or be construed as libelous. Care should also be taken that no information is disclosed that could have an adverse economic impact on the organization or provide a competitor with an advantage.

The research manager can help his employees get published by watching all "Call for Papers" and submission requests found in the professional journals that he scans. These should be routed to people in his unit who show promise and could reflect positively on the organization. Each organization should have a policy stating the requirements for review and approval prior to release of all publications. By alerting the employees to potential papers, instructions and precautions can be emphasized before an article is written. This can make the job of reviewing papers easier for the supervisor.

When the decision is reached to publish an article, aside from selecting the topic, the potential author must take note of critical dates (abstract due and manuscript due); required format for the abstract, manuscript, illustrations, tables, scientific notation, and references; cost of preparation; possible costs of publication and reprints; and to whom the abstract must be submitted for approval (Day 1979). Great care should be taken in selecting a title, as it will be read far more often than the paper and will appear in many locations with the writer's name (in reviews, bibliographies, reprints, abstracts, and curricula vitae). The abstract should be succinct and prepared according to the publisher's instructions, and it should clearly explain the purpose of the paper. The effectiveness of the abstract will determine whether a reader wants to continue through the entire paper. In tackling the paper itself, it is important to follow the rules of good writing, obey the publisher's manuscript guide, and heed the advice of the editor.

PRESENTING SEMINARS

Professional presentations are useful for helping your employees develop speaking skills and other talents that will make them a greater asset in your department and the organization. Employees frequently request permission to attend meetings so they can further their technical skills. A smart supervisor will attempt to link meeting attendance with presenting a paper, whenever this is practical. In this way, while the organization pays for the trip and the individual gains from the experience and the interaction, the employee must practice his skills and the organization receives some valuable publicity. The supervisor should start this policy with his younger staff early in their careers, so that their skills are refined as they move into more senior positions.

Not everyone can get a paper on the schedule of a national, international, or sectional meeting. In addition, many employees will feel very uncomfortable speaking before a large group of strangers from varying backgrounds. The laboratory manager can help his employees by having them make presentations among the members of the department or research team. With friends and associates with whom he works daily, an individual who might be nervous in front of an impersonal gathering will be able to develop the confidence he needs to speak before a group. Practice is the key to a relaxed speech and except for those few individuals with a natural delivery, almost everyone needs advance preparation and lots of practice.

Since most talks are only ten to thirty minutes long, the speaker must make every effort to get his message across quickly. Time does not permit freedom to ramble over various topics. There is time to present only one or, in some cases, just a very few limited ideas. In most oral communication, points are repeated in order to ensure that the listener has gotten the message. This is also true in presenting a seminar, where the speaker should (1) use the introduction to tell the audience what they are going to hear, (2) tell it to them in the body of the talk, and (3) during the conclusion tell them what they should have heard.

The skillful speaker will attempt to be as effective as possible in the limited time provided (Lewis 1982). There are several ways to do that.

- Identify the topic and be sure you are qualified to present it.
- Get to the point quickly and say what has been promised.
- Be well prepared and do not wait until the last minute to write the talk.
- Just prior to going on the platform, walk around and take some deep breaths (this level of exercise can help with nervousness or just improve a relaxed delivery).

If you are prepared and know your material, then there is no reason to be nervous and your talk stands every chance of being both effective and successful.

The method of delivery will affect how a speech is received. There are no absolute rules that will guarantee a successful outcome. Each speaker must develop his own criteria with time and practice. The best talks are not read, but spoken from memory or extemporaneously. However, many employees are not "naturals" and may benefit from some hints that have helped other neophyte speakers.

- Have your manuscript fully typed and in front of you, even if you decide not to read it.
- Be sure the text can be easily seen at the podium, so you do not have to search for a word or idea.
- Make frequent eye contact with the audience, as this will improve your credibility.
- Pick a real or imaginary person in the back of the room and speak in a loud clear voice, as if he could have a hearing problem.
- Speak more slowly than usual and use pauses for emphasis.
- Repeat a critical point, if it is possible someone might not have understood it.
- When pointing to visual aids never turn your back on your audience.

Above all else, speak with confidence and present a positive attitude, because you are representing your profession and your employer.

GRAPHICS AND AUDIOVISUAL MATERIALS

As an old Chinese proverb says, a picture is worth a thousand words. The same thing applies to technical presentations. Graphics can be used to help get the message across and present data in a clear and concise way, as well as to hold the interest of the audience. Charts, chalk or marker boards, flip charts, transparencies, and slides are some of the commonly used visual aids, and each has its strong points as well as inherent weaknesses. The type of visual aid selected should be appropriate to the purpose for which it is used.

Flip charts probably produce the best effect in spontaneous meetings in which the spokesman is trying to capture ideas or expand on a topic. A disadvantage for the spokesman who does not have a clear handwriting or is not an artist is that the visual portion may be illegible or communicate poorly, reducing the effectiveness of the presentation. It may help to try to improve poor handwriting or prepare the charts ahead of time. If the speaker wishes to retain a degree of spontaneity, some of the graphics and selected key words could be prepared ahead of time and the details filled in during the talk. Another approach is to sketch the information in lightly before the meeting, so that during the presentation, the speaker need only go over the material with markers. From any distance, the audience will never know the difference.

Charts should always be prepared ahead of time, and if the speaker is not artistically inclined, it is best to have someone in a graphics department prepare the material. When professionally prepared using color, large distinct lettering, and clear artwork, their impact on the audience can be strong.

The use of transparencies is a very economical system of presenting graphics. The material can be prepared free hand or typewritten. The speaker can control eye contact and improve audience attention by the timing in which the transparencies are placed or removed and by turning off the overhead light. An additional advantage of the transparency is that during a discussion, a grease pen can be used to write on it to highlight certain values or the critical points of a curve.

The 35-millimeter slide gives a presentation a professional appearance but is more expensive than the other visual aids that have been mentioned. If photographic services are available, materials that have been prepared by a secretary can be easily turned into slides even using varying colors for the background or lettering to add emphasis. Besides making the presentation more costly, this method requires more lead time, lacks spontaneity, and does not allow for last-minute changes.

Computer graphics, in the form of output for handouts, transparencies, or 35-millimeter slides, are beginning to become popular as a medium for the speaker. Direct output can be used as well, provided the output screen is large enough for the audience to see. Keep in mind how large the room is, how many will be in the audience, and what it is you want them to see. Most screens are too small for the purpose. Even with large screens, details can be seen only in a small room. Projection screens are required for larger rooms. For most seminar rooms, the audience will be best served by plotting transparencies from your screen output. This assures quality and detail.

Other presentation systems include the dissolving of multiple slide programs with a sound cassette, movies, and video cassettes. As these methods are more specialized and require special equipment and training to produce, they will not be covered here. A local graphics or audiovisual studio can provide the necessary assistance if one of these methods would be useful.

In preparing a talk, the speaker should understand the purpose of graphics, which fall into two principal categories: to convey information and to hold interest. The data or results of several months' work could easily fill an entire volume and need to be reduced to essentials in making a presentation. A clear graph can help get the speaker's message across while a transparency full of numbers can quickly lose the audience's interest. When preparing a graph, be sure that it communicates what you are trying to say as simply and effectively as possible. Where your data might indicate five or six different results, the graph may be complicated and difficult for someone viewing it for the first time to understand. Be sure it shows only the information that is necessary to the point you are making. If you need to cover several points, consider whether it might be better for the audience to see several different graphs. Another technique is using separate transparencies showing each curve in a different color, so that each element appears at a time. By laying them over the graphs, the speaker can illustrate what is happening with each set of data and compare the points of interest on the different curves. With all graphics, but especially when more than one piece of information appears on the same graph, be sure that all points, curves, scales, and axes are clearly labeled.

A powerful but often overlooked tool for presentations (Shulman 1966), graphics can be used to get or to hold the attention of the audience, and to recapture the interest of those whose minds might have wandered. A humorous sketch or a cartoon can be shown briefly to establish or reestablish eye contact. This is very effective in opening a talk and in summarizing a talk, but it can also be useful during a talk—especially if the presentation is long. Think of your audience. If you are not the only speaker, and people have been listening for over an hour, their attention may be fading. Look for a place in the talk where some levity can be introduced, such as after an intensive examination of data. A good speaker may key several points for humor. By reading the audience's mood, he will know whether to use or to skip over the anecdotes or cartoons he has prepared.

HINTS FOR BETTER TECHNICAL WRITING

The purpose of this section is to suggest a few simple rules to follow when preparing a report. The selected bibliography at the end of this chapter provides more detailed information for managers seeking to improve their own skills or those of their employees. References specifically designed to help with technical communications (for example, Tichy 1966, and Day 1979) have been included, together with resources relating to general, business, or specialized communications.

Here are a few general rules to help you.

- *Do not put it off.* Time is money and the longer you delay in producing the report, the more it will cost your organization.
- *Plan ahead before you start writing.* Think your topic through. Prepare an outline. Prepare a draft, let it cool, and then go back and revise it.
- *Know your audience.* Be aware of your audience. Imagine that you are writing for a particular individual whom you know well. This device usually helps the writer present the material clearly and without excessive technical jargon.
- *Keep it clear.* Make sure that any technical terminology used will be understood by the reader and supply definitions where needed.
- *Be thorough.* Make sure all aspects of your topic have been covered. Use appendices as necessary to provide backup material, but do not force your reader to come back for clarification.
- *Stick to your topic.* Know what the single message is and say it without getting on to other subjects. If there are side issues in the paper, it was probably not adequately defined.
- *Keep it simple.* Do not write to impress, because that will turn your audience off. Showing off one's vocabulary or intellect in a paper does not produce a paper that will make a lasting impression.
- *Be honest.* The introduction, abstract, or foreword should accurately reflect the subject and the message. Do not deceive your audience.
- *Be brief.* Do not waste the valuable time of your audience.

As you develop your skills, you will learn how to use emphasis when it is needed to make a point and how to improve your style. Technical reports do not have to be dull or boring. One final thought: if you put yourself in the reader's shoes when you review your paper, you will become aware of anything that keeps you from getting your message across.

COMMUNICATING TECHNICAL RESULTS TO THE LAYMAN

In the language of the sciences the possibilities of interpretation and the need for definition require close attention, especially in addressing an audience that is unfamiliar with scientific concepts and terminology. When communicating with the public, it is the duty of anyone using a specialized vocabulary to interpret his message in the common vernacular (Lewis 1982). Failure to do this often leads to widespread confusion as well as a general distrust of the individual, the profession, or the organization.

Every specialized group develops a language of its own in order to communicate effectively. This in itself has caused difficulty between groups such as manufacturing and research, or between research and the nontechnical departments. Language barriers exist because of both the technical nature of the subjects and the specialized words and phrases that are used. Personnel in areas such as manufacturing and marketing generally learn enough about each other's field to allow adequate communication. In most cases, however, there appears to be a special barrier when it comes to research.

Since it is not feasible for nontechnical personnel in other departments to learn the specialized terminology, it is incumbent on research personnel to ensure that all external communications be translated into words that others can understand. Research takes place within the context of a business, a government agency, a university, or in a social environment; therefore, technical personnel must be capable of communicating in these settings. In most cases, research decisions will be made in other departments, and researchers will have a much easier time if they work cooperatively and do not create an atmosphere that places them in an adversarial role.

THE OBLIGATION OF THE TECHNICAL SPOKESMAN TO THE PUBLIC

Every person in every department in every company relates to the general public; and managers must do what is necessary to ensure that their personnel can handle their unofficial public relations role (Furnas 1948). This represents a special problem for technical personnel because of the inherent difficulty of understanding their fields. Where public relations specialists might communicate to the general public in business or legal areas, many times they defer to one of the senior technical specialists to communicate a complex scientific topic. A laboratory member might suddenly find himself thrust in the role of communicating to the media. At the very least, he may have to make his specialty (for which he has spent many years training) understandable to the public relations specialist, when there are legal or particularly sensitive issues involved.

The public relations role extends beyond the workplace to each employee's communications with friends, neighbors, and associates. It is important to understand that speaking out in a technical area without being sure that the meaning will be understood could lead to a public outcry. The role of public relations does not belong exclusively to one person or one department, but can fall on every technical specialist. The laboratory manager should be sure that the people in his unit are thoroughly familiar with legal and social issues that may relate to their laboratory activities. Everyone must be familiar with the company's policies with respect to any sensitive areas of research. Employees must be cognizant of what they say to the people around them to ensure that no misunderstanding can take place, for a confusing bit of jargon could come back to haunt the company.

Some important principles to remember in meeting obligations to the public include the following:

- Keep the door open, as few things terrify the nontechnical person more than the cloistered, secret attitude portrayed in many movies about scientists.
- Understand your company's policies, and be sure every employee understands the company's policies, as well as the related social issues.
- Explain to the employees that they are ambassadors from research to other parts of the company, to the government, or to the public at large, and they should be careful in what they say and how they say it.

Assuming the company has a public relations department, be sure that employees do not speak with newspaper and other public media without first consulting the research department head and the public relations department. The intention is not to hide something, but rather to ensure that media go through the appropriate department or chain of command to obtain information.

Remember that your company, your department, and your personnel are members of a community, and it is important to maintain a good working relationship with the community. As long as the people in the surrounding areas feel that they receive cooperation and there is an open, positive attitude, then it is far less likely that they will attempt to take legal action concerning activities by the company. There are a number of ways in which technical personnel can help foster that positive working attitude.

- Keep the lab environment clean, as the creation of dust and odors or an unkempt and messy exterior can be an irritant that might help set off community reaction.
- Play to the community's desire to know what is happening by

offering plant tours and open houses, within the guidelines of the company or university.
- Provide technical assistance to the local government, instruction at local colleges, demonstrations for school classes, or assistance on career nights in the local school system.
- Have qualified and knowledgeable people talk to the local service and social clubs.

A little effort by a few people can go a long way towards developing effective communications and a positive attitude with the community and people around you.

SUMMARY

Communications are the life's blood of an organization and of all technical fields. Without a healthy exchange of accurate and clear ideas, no organization can be productive and no field can make progress. Very often, for people going into engineering and the sciences, the communicative arts and technology are considered incompatible. Whether this is due to the nature of the educational process or some inbred bias of society, scientists must learn that communications and technology are not vastly separated fields. Rather, they are inseparable if one wants to be successful. As great an effort must be put forth in learning to be good at communicating with others, as was in developing the technical skills. If formal education did not adequately prepare the scientist for his job, he must seek the needed skills through continuing education. A specific task of the good manager is to help his employees develop the necessary skills in writing and understanding what is written, in speaking as well as listening, and conducting or participating in an effective meeting. The research manager that meets his obligations for good communications in and from his department is doing his part to make his organization as productive as possible.

REFERENCES

Day, R. A. 1979. *How to Write and Publish a Scientific Paper.* Philadelphia: Institute for Scientific Information.
Furnas, C. C., ed. 1948. *Research in Industry.* Princeton: Van Nostrand.
Lewis, D. V. 1982. *Secrets of Successful Writing, Speaking and Listening.* New York: AMACOM (a division of the American Management Association).

Shulman, J. J. 1966. *Effective Writing for Engineers, Managers, and Scientists.* New York: John Wiley.
Tichy, H. 1966. *Effective Writing for Engineers, Managers, and Scientists.* New York: John Wiley.

GENERAL BIBLIOGRAPHY

On Technical Communications Skills

American Chemical Society. 1978. *Handbook for Authors.* Washington, D.C.: ACS.
———. 1979. *Manuscript Requirements.* Washington, D.C.: ACS.
Ammon-Wexler, J., and C. Carmel. 1976. *How to Create a Winning Proposal.* London: Mercury Communications.
Brogan, J. 1973. *Clear Technical Writing.* New York: McGraw-Hill.
Day, R. A. 1979. *How to Write and Publish a Scientific Paper.* Philadelphia: Institute for Scientific Information.
Ewing, D. 1979. *Writing for Results in Business, Government, the Sciences and the Professions.* 2d ed. New York: John Wiley.
Freedman, G. 1967. *A Handbook for the Technical and Scientific Secretary.* New York: Dover.
Jones, F. D. 1931. *How to Write a Technical Article.* New York: Industrial Press.
Jordan, S., J. M. Kleinman, and H. L. Shimberg, eds. 1971. *Handbook of Technical Writing Practices.* New York: Wiley-Interscience.
Lefferts, R. 1978. *Getting a Grant.* Englewood Cliffs, N.J.: Prentice-Hall.
Mills, G., and J. Walter. 1970. *Technical Writing.* 3d ed. New York: Holt, Rinehart & Winston.
Norgard, M. 1959. *A Technical Writer's Handbook.* New York: Harper & Row.
O'Connor, M. 1979. *The Scientist as Editor.* London: Pitman Medical Publishing.
O'Connor, M., and F. P. Woodford. 1978. *Writing Scientific Papers in English.* London: Pitman Medical Publishing.
Rathbone, R. R. 1966. *Communicating Technical Information.* Reading, Mass.: Addison-Wesley.
Sherman, T., and S. Johnson. 1975. *Modern Technical Writing.* 3d ed. Englewood Cliffs, N.J.: Prentice-Hall.
Smith, R. W. 1963. *Technical Writing.* New York: Barnes & Noble.
Society for Technical Communication. 1973. *Proposals... and Their Preparation.* Distributed by UNIVELT, San Diego, Calif.: Washington, D.C.
Tichy, H. 1966. *Effective Writing for Engineers, Managers, and Scientists.* New York: John Wiley.
Ulman, J. N., and J. R. Gould. 1972. *Technical Reporting.* New York: Holt, Rinehart, & Winston.
Weisman, H. *Technical Report Writing.* 1966. Columbus, Ohio: Merrill.
Weiss, E. H. 1982. *The Writing System for Engineers and Scientists.* Englewood Cliffs, N.J.: Prentice-Hall.

On General Communications Skills

Barzun, J. *Simple and Direct.* 1976. New York: Harper & Row.
Bernstein, T. 1965. *The Careful Writer.* New York: Atheneum.
_____. 1977. *Reverse Dictionary.* New York: Quadrangle.
Berry, T. E. 1971. *The Most Common Mistakes in English Usage.* New York: McGraw-Hill.
Blumenthal, L. 1976. *Successful Business Writing.* New York: Grosset & Dunlap.
Corbett, E. P. J. 1973. *The Little English Handbook: Choices and Conventions.* New York: John Wiley.
Eckersley-Johnson, A. L., ed. 1976. *Webster's Secretarial Handbook.* Springfield, Mass.: Merriam.
Ellenbogen, A. 1978. *Letter Perfect.* New York: Macmillan.
Fearnside, W., and W. Holther. 1959. *Fallacy: The Counterfeit of Argument.* Englewood Cliffs, N.J.: Prentice-Hall.
Follet, W. 1966. *Modern American Usage.* New York: Hill & Wang.
Goeller, C. 1974. *Writing to Communicate.* New York: New American Library.
Irmscher, W. 1972. *The Holt Guide to English.* New York: Holt, Rinehart & Winston.
Janis, H., and H. Dressner. 1972. *Business Writing.* New York: Barnes & Noble.
Lewis, D. V. 1982. *Secrets of Successful Writing, Speaking & Listening.* New York: AMACOM (a division of the American Management Association).
Maslow, A., ed. 1959. *New Knowledge in Human Values.* New York: Harper & Row.
Mitchell, R. 1979. *Less Than Words Can Say.* Boston: Little Brown.
Newman, E. 1980. *On Language.* New York: Warner Books.
Perelman, C., and L. Olbrechts-Tyteca. 1969. *The New Rhetoric: A Treatise on Argumentation.* Notre Dame, Ind.: University of Notre Dame Press.
Shaw, H. 1970. *Errors in English and Ways to Correct Them.* 2d ed. New York: Barnes & Noble.
Strunk, W., Jr., and E. B. White. 1979. *The Elements of Style,* 3d ed. New York: Macmillan.
Toulmin, S. 1958. *The Uses of Argument.* Cambridge, England: Cambridge University Press.
Urdang, L., ed. 1972. *The New York Times Everyday Reader's Dictionary of Misunderstood, Misused, Mispronounced Words.* New York: New York Times Book Co.
U.S. Government Printing Office Style Manual. 1973. Rev. ed. Washington, D.C.: Government Printing Office.
Vardaman, G., and P. B. Vardaman. 1973. *Communication in Modern Organizations.* New York: John Wiley.
Waddell, M., R. Esch, and R. Walker. 1972. *The Art of Styling Sentences.* Woodbury, N.Y.: Barron's Educational Series.
Windes, R., and A. Hastings. 1969. *Argumentation and Advocacy.* New York: Random House.
Zinsser, W. 1980. *On Writing Well.* 2d ed. New York: Harper & Row.

Chapter 6
Patents and Proprietary Information Management

PROPRIETARY INFORMATION

The term "proprietary information" as used in this chapter refers to information owned by the particular research organization responsible for its generation. Conversely, information that is not so owned is in the public domain and is therefore not proprietary. Since the possession of proprietary information can provide an advantage in the marketplace, the desirability of maintaining the exclusive right to its use is apparent. The owner of proprietary information has a protectable property right in it. This information is sometimes referred to as intellectual property to distinguish it from tangible personal property. Unlike the normal situation with tangible property, the owner of intellectual property must take care that his ownership is not lost. The inventor of a new machine can leave it in a place freely accessible to the public without fear that someone else can lawfully acquire the machine itself for private gain since our courts would not sanction the conversion of this tangible personal property. However, the intellectual property—that is, the proprietary information, which is revealed by a view of the machine can be freely used by anyone who so desires in the absence of an affirmative act by the owner to protect his right of exclusivity. This principle was emphasized by the United States Supreme Court in 1964, when it held that in the absence of patent or copyright protection Sears, Roebuck & Company had every

---

This chapter was written by Jerome L. Jeffers, J.D., Associate Patent Counsel, Miles Laboratories, Inc., Elkhart, Indiana.

right to make and sell copies of a pole lamp that had been marketed by the Stiffel Lamp Company. The general rule concerning intellectual property is that once it enters the public domain it belongs to the public.

If the fruits of a research organization's efforts are to be maintained as proprietary information and thereby provide the exclusive position in the marketplace that is often necessary to justify the cost of the research, care must be taken to protect this intellectual property. Two ways of providing this protection that are of interest to the laboratory manager are maintenance of trade secrets and the procurement of patents.

TRADE SECRETS

Several years after the decision in the Sears pole lamp controversy, certain employees of the Kewanee Oil Company misconstrued the Supreme Court's opinion as holding that trade secret protection had been preempted by the patent laws. When these employees departed to form their own company and compete with Kewanee by using proprietary information garnered during their term of employment, they defended a suit for misappropriation of trade secrets by arguing that under the Sears decision they had not acted improperly. Fortunately, the Supreme Court decided in favor of the former employer. This decision was based on the finding that the technology in question, a method of growing synthetic crystals useful in the detection of ionizing radiation, was not apparent from examination of the crystals themselves, which were available to the public without restriction. The fact that the former employees learned the crystal-growing technique during their term of employment placed them in a confidential relationship with their former employer. By condemning this misuse of information obtained pursuant to the employer/employee relationship, the Supreme Court gave a clear signal that trade secret law is alive and well in the United States.

A trade secret has been defined as

> Any formula, pattern, device or compilation of information which is used in one's business and which gives him an opportunity to obtain an advantage over competitors who do not know or use it.

This definition covers patentable inventions but also relates to information that does not meet the rigorous standards of patentability. In some cases it may be desirable to forego patent protection in favor of maintaining a trade secret even in the case of patentable subject matter. This decision, which is typically made at the top levels of the organization sponsoring the research, involves an analysis of such factors as the ability to keep the technology secret and to police any patents covering the invention. For example, an improvement in process technology might be clearly patentable but can be practiced in secret by its inventor. Furthermore, publication of this sort of invention in the patent literature may be undesirable due to the difficulty in detecting infringement. These considerations may result in a decision to retain the right to the exclusive use of the technology by keeping it secret rather than seeking patent protection. Once the decision to rely on trade secret protection is made, care must be taken to not disclose the information except under an implied or (preferably) express obligation by the receiving party not to disclose or use it. The typical employee agreement will expressly create this obligation when entered into by employees. The former Kewanee Oil Company employees had entered into such an agreement, so it is an open question as to whether the Court would have found an implied obligation in the absence of the employee agreement; this would depend on the totality of the circumstances of the relationship. Our courts have been reluctant to place too many restrictions on the right of former employees to use their skills either as entrepreneurs or in the service of a new employer, but they will not tolerate outright theft as in the Kewanee case. The trend has been to attempt to strike a balance between the employee's right to earn a living and the former employer's right to have its trade secrets maintained in confidence.

Disclosure of trade secrets to others in order to achieve business or research objectives should only be done under an express agreement of confidentiality. A word of caution is appropriate at this point regarding the receipt of confidential information from others. Once the information is received, it cannot be used or disclosed by the receiving party; and while information known by the receiving party at the time of its receipt will normally be excluded from the obligation, few confidentiality agreements exclude information that is independently developed by the receiving party after its receipt from the disclosing party. The potential detrimental effects of receiving confidential information relating to an area of ongoing or anticipated research activity is apparent. When information is to be received on a nonconfidential basis, it is desirable to enter into a written agreement to this effect with the disclosing party. Such an agreement will eliminate any potential finding of an implied obligation of confidentiality as well as prevent any misunderstanding between the parties as to the terms of disclosure.

Protection of proprietary information by maintaining it as a trade secret has certain advantages over patent protection. For one thing, the duration of this form of protection is for as long as the secret can be maintained, which may be longer than the seventeen-year lifetime of a United States patent. Furthermore, certain types of proprietary information such as computer software are not regarded as patentable subject matter under our laws. On the other hand, one who independently invents the subject matter of the trade secret has every right to use it in competition with the first inventor. It has been argued that the second inventor could obtain patent protection and exclude the first from continuing to use the invention, although our law is not entirely clear on this issue. Another disadvantage of the reliance on trade secrets is that, like Sears in the pole lamp controversy, anyone who elucidates the secret by legitimate means is free to practice it. The Stiffel pole lamp was simply copied. In some cases reverse engineering may be required, and this is not unlawful in the absence of a confidential relationship. Reverse engineering involves looking at a product to elucidate how it was made. To the extent that this can be done, "secret" manufacturing procedures lose their status as secret once a product is marketed. The method of growing the synthetic crystals in the Kewanee Oil case could not be determined from an examination of the crystal, so Kewanee had an enforceable property right in the secret crystal-growing method. While the acquisition of trade secrets through breach of a confidential relationship is unlawful (as was reaffirmed in the Kewanee Oil case), the facts necessary to prevail in a trade secret misappropriation case can be difficult to prove and protracted litigation may result.

PATENTS

Recourse to patent protection can obviate some of the disadvantages inherent in reliance on trade secret law to preserve the exclusivity inherent in the ownership of proprietary information. Our patent law finds its basis in Article I, Section 8, Clause 8 of the U.S. Constitution:

> Congress shall have the power ... *To promote the progress of* Science and *useful arts, by securing for limited times to* authors and *inventors the exclusive right to their* respective writings and *discoveries.*

This constitutional provision is intended to provide a foundation for both the patent and copyright laws; the italicized portions relate only to patents. The word "science" is not emphasized because in eighteenth-century parlance it meant all knowledge and hence applied to the copyright provision. Applied science (useful arts) is what the founding fathers had in mind in providing for a patent system. The useful arts include ornamental designs since design patents are obtainable under our patent system. In addition, some ornamental designs may be appropriate subject matter for copyright protection. These considerations are beyond the scope of this discussion, which is limited to utility inventions—that is, those that are useful for reasons other than their aesthetics.

One way in which the patent system promotes the progress of the useful arts is by providing a period of exclusivity during which the patent's owner can profit from his invention without the threat of competitive price cutting. The object here is not necessarily to make patentees rich but to hold out the prospect of enhanced profitability to encourage the expenditures required to invent new and useful processes and products. This stimulus may be of greater importance to the employed inventor than to the individual because company financial officers tend to analyze the bottom line, whereas the individual inventor is often driven by other motivating factors. By being ever conscious of patent considerations, the laboratory manager can help provide his employer with a reasonable return on investments in research and development.

Patent protection, or the lack of it, can also be a fundamental element in the decision on whether to invest the resources necessary to bring the invention from the laboratory to the marketplace. Chester Carlson patented xerography (or electrophotography as he called it) in the 1940s, but the Xerox 914 copier did not reach the market until 1960. The large expenditure of time and money required to bring Carlson's laboratory curiosity to the point of commercial feasibility was justified by the prospect of marketing exclusivity and the high profit margins inherent in the sole rights to a product the public desires. The government policy relating to ownership of patent rights to the fruits of federally funded research has changed in recognition of the necessity for economic incentive if inventions are to be commercialized. Until recently the typical government R & D contract provided for patent ownership by the federal agency providing the funding with a nonexclusive license to the contractor. In the absence of the incentive offered by marketing exclusivity, few inventions were developed for nongovernmental purposes. This problem has been partially ameliorated by an act of Congress providing for ownership of patent rights resulting from federally funded research under certain circumstances. In the interest of fairness to the taxpayer, the funding agency retains a nonexclusive license to use the invention for governmental purposes and has march-in rights in the event the contractor fails to commercialize the invention within a reasonable time.

Patentable Subject Matter

Assuming a decision is made that patent protection is desired for a particular invention, it becomes necessary to determine if such protection can be obtained under the prevailing circumstances. Because of the unique considerations involved, this determination should be made only after consultation with an experienced patent attorney. The first consideration in assessing whether an invention is patentable is to determine whether it falls within the definition of a process, machine, manufacture, composition of matter or improvement thereof, since these are the types of subject matter for which Congress has provided the availability of utility patent protection. Our courts have consistently held that a natural law such as the force of gravity or the formula $E=mc^2$ cannot be patented under the congressional definition of patentable subject matter. (One might question the constitutionality of any act of Congress that purported to render patentable a law of nature, although the courts have not yet found it necessary to rely on constitutional considerations in this regard.) Products of nature, such as chemical entities found in animal and vegetable life, are also considered unpatentable with the proviso that a naturally occurring chemical that requires extraction and/or purification in order to render it suitable for use can be regarded as patentable subject matter if the naturally occurring material was not in a form that was useful to mankind. Acetylsalicylic acid (aspirin) was a molecule found in nature and yet its pure form suitable for ingestion as an analgesic and anti-inflammatory agent was held to be patentable in the famous *Farbenfabriken of Elberfeld Company* case.

A milestone of sorts was reached in 1980 when the U.S. Supreme Court rendered its decision in the much publicized case of *Diamond v. Charkrabarty*. This controversy involved the refusal of the U.S. Patent and Trademark Office (through its Commissioner Diamond) to grant Charkrabarty a patent on his admittedly new and useful bacterium. This refusal was based on the commissioner's belief that Congress had not intended living matter to be the subject of the patent grant. The Supreme Court found by a narrow (five to four) majority that Congress intended patentable subject matter to include anything under the sun that is made by man. In deciding that a living organism could be patented under the existing patent statute and without the need for special legislation, the Court did not let down all barriers for it reiterated its holding in an earlier case that an improved method of calculation (that is, one carried out using a preprogrammed computer) was not patentable subject matter even when the calculation was tied to a specific end use. Some confusion was caused in 1981 when the Supreme Court held that a rubber-curing system that employed a preprogrammed computer continuously to solve the Arrhenius equation, with temperature being the variable, was patentable subject matter. Suffice it to say that the use of a preprogrammed computer in a process does not necessarily render the process unpatentable, and some computer control schemes may be regarded as patentable. Certainly the decision in the Charkrabarty case suggests that a majority of the Supreme Court takes an expansive view of those technologies that qualify for patent protection and will not insist that Congress rewrite the patent statute every time an unforeseen technology such as genetic engineering evolves. In the Charkrabarty decision, Chief Justice Burger commented on the incentive provided by the patent system as follows:

> Whether respondent's [Charkrabarty's] claims are patentable may determine whether research efforts are accelerated by the hope of reward or slowed by want of incentives ... (*Diamond v. Charkrabarty*)

Utility, Novelty, and Unobviousness

Once the determination is made that a particular invention is of the type that conforms to the statutory concept of patentable subject matter, the questions of whether it is useful, new, and unobvious must be considered in determining its patentability. Utility is not a serious barrier in the normal case because our laws require only that the invention perform a meaningful function, not that it operate in a manner superior to what has gone before. The Supreme Court has held, however, that the usefulness requirement is not satisfied in the case of a process for the synthesis of a compound whose only known utility was that it belonged to a class of compounds that was the subject of serious scientific investigation. Since the method of preparing a compound useful only in research is not patentable, a fortiori, the compound itself would not satisfy the utility requirement. Conversely, a novel analytical method would be considered useful since, while it does not produce a product, it is a process that provides a useful result.

In order to be patentable the invention must not only be useful; it must be novel. This sounds axiomatic, but for the sake of fairness, our patent statute defines novelty in a very precise manner. To begin with, the statute states that "A person shall be entitled to a patent unless—" and then goes on to set out conditions that will defeat the novelty of an invention. The preamble is noteworthy because it places the burden of proof on one who would defeat novelty rather than on the inventor to establish novelty.

Under the statutory scheme an invention is not novel if it was known or used by others in this country or described in a printed publication anywhere in the world before its invention by the applicant. Prior invention by another will not defeat patentability by a later inventor if the prior inventor does not publicly use or disclose the invention in the United States or describe it in a printed publication. Note that use or disclosure (short of a description in a printed publication) outside of this country is not a bar to patentability. This is not to say that one could observe an invention (which had not been described in a printed publication) in a foreign country and then obtain a valid United States patent. While our statute does not require that the patentee be the first inventor, it does require that he be an inventor in fact. One who derives an invention from another is not an inventor.

Another circumstance that will defeat novelty is where the invention was in public use or on sale in this country or described in a printed publication anywhere in the world more than one year before the filing date of the application for patent. Accordingly, when an inventor is apprised of the divulgence of his invention by another, he must take steps to have his patent application filed no more than one year from the date of such disclosure. An inventor can bar himself from obtaining a patent by a premature disclosure, sale, or publication of the invention; and if such an event takes place, it is essential that the patent application be filed no more than one year later. If foreign patent protection, particularly in the western European countries, is desired, one cannot even rely on the one-year grace period provided by our statute. This is because the European Patent Convention, to which most of the economically important European countries are signatories, subscribes to the concept of absolute novelty. Under this concept, any disclosure of the invention anywhere in the world by whatever means will immediately defeat its patentability. For the reasons set out above, it is necessary for the laboratory manager to deter the premature disclosure of inventions for which patent protection is deemed desirable.

The United States patent statutes also provide for denial of patent protection to one who has abandoned the invention for which such protection is sought. The express abandonment of a commercially

tion before the PTO. This holding can be regarded as punitive because there was no finding that the patent application would not have been allowed had all of the data (favorable and unfavorable) been included in the affidavit.

The lesson to be learned by the 3,4-DCPA/3,4-DCAA controversy is that the utmost candor must be used when dealing with the PTO. Patent examiners are experts in their field and the courts will give great deference to their actions if such actions are based on all the material facts. However, when the applicant fails to bring all such facts to the attention of the examiner, the decision to allow the application and issue the patent is seriously flawed. The laboratory manager must be very cognizant of this duty of candor when dealing with his patent attorney, so that the attorney can bring all material considerations to the attention of the PTO. This duty also extends to the citation of prior art during the patent solicitation process. The examiner will conduct an independent search, but time and resource limitations as well as lack of investigatory capability can result in material prior art being missed. Accordingly, the applicant for patent is under a duty to disclose all prior art that is material to the examination of the patent application. All prior art includes not only printed publications but also use or sale of the invention as those terms have been defined earlier in this chapter.

Another way in which some objectivity can be applied to the question of unobviousness involves what the Supreme Court referred to as secondary considerations in a decision rendered in 1966. These secondary considerations include long felt but unsolved need, failure of others to solve the problem, and commercial success of the invention. An invention that looks obvious in hindsight may take on a different light if it can be shown that there was a long felt need for the invention that was not fulfilled. Likewise, the failure of others to solve the problem and/or the commercial success of the invention can tilt a close case toward a finding of unobviousness. For a showing of failure by others to carry much weight it must be shown that the "others" had the capability and inclination to make the invention if it were in fact obvious. Commercial success must be a result of the merits of the invention, and its effectiveness will be diminished by a showing that extensive advertising or other promotional schemes were used to enhance the commercialization of the invention. While these "secondary" indicators of unobviousness will not apply in very many cases, they should be kept in mind because when they do apply the invention will tend to be one of commercial significance.

Inventorship

No discussion of United States patent practice would be complete without emphasis being placed on the proper determination of inventorship. Our practice is somewhat unique in that it requires that only the actual inventor(s) apply for the patent. To be an actual inventor, one must have made an *unobvious* contribution to the invention as claimed. The emphasis is on the word *unobvious* because many meaningful contributions involved in making the invention are not unobvious and therefore do not qualify the one making them as an inventor. For example, a laboratory technician who carries out the manipulative steps necessary to synthesize a novel compound at the behest of a supervisor would not be regarded as a coinventor. This is because the supervisor would have scoped out the synthesis in advance and no unobvious act can be found in using standard chemical techniques to prepare the compound. Conversely, if the laboratory technician were to discover an unexpected parameter necessary to make the reaction go during the synthesis, he could be considered an inventor at least as to the process. The laboratory manager who makes suggestions during the course of research conducted by his staff may or may not be an inventor depending on the nature of the suggestions. The proper determination of inventorship is a difficult, but necessary, task if enforceable patents are to be obtained. The patent attorney is often faced with the onerous task of deciding that an individual who contributed to making an invention is not a coinventor because the contribution was not inventive—that is, not unobvious. The laboratory manager can bolster morale in these situations by devising alternative methods for recognizing the contributing noninventor. An incorrect determination of inventorship is correctable if it can be shown that an honest mistake was made, but the process of correction is difficult and time consuming. An uncorrected or uncorrectable determination of inventorship can result in invalidation of the patent, and it should be remembered that naming someone as an inventor who is not actually an inventor is just as serious an error as failing to name someone who made an inventive contribution.

The Patent Application

Once the patent attorney, after consultations with the research personnel and review of the prior art, has decided that the invention constitutes patentable subject matter that is believed to be useful, novel, and unobvious and the inventorship has been determined, the patent application can be prepared and filed with the PTO. The application will consist of a specification and claims, with drawings included if they are required to provide full understanding of the invention. The purpose of the claims is to circumscribe precisely what the applicant regards as his invention. This facilitates the determination as to whether the invention, as claimed, is patentable; and when the patent is issued, the precision required for proper claim drafting will apprise others as to what the patentee regards as his area of exclusivity. The purpose of the specification is to describe the invention in such a manner that anyone skilled in the art will be taught how to practice it. Our patent laws also require that the applicant disclose the best mode contemplated for practicing the invention. This is an important consideration since it prohibits the seeking of patent protection while keeping the preferred method of practicing the invention as a trade secret. The decision to rely on trade secret or patent protection must be clear-cut since our laws do not provide for a hybrid type of protection. The difficulty in making this decision is ameliorated by the fact that U.S. patent applications are kept under strict secrecy until the patent is issued. Accordingly, it is possible to file a patent application in this country and let it become abandoned if the decision to rely on a trade secret rather than patent protection is made after filing of the application but before its issuance. The requirement that the inventor disclose the best mode refers to the best mode at the time of filing; there is no requirement that the specification be updated to include post-filing improvements to the invention. Legislation that would have required such an updating was passed by the Senate in the mid-1970s but never came up for a vote in the House of Representatives.

When the patent application is filed, it will be assigned to an examiner who has gained expertise in the field of technology to which the invention relates. When the examiner gets to the application in the ordinary course of business, he will respond with an official office action. Since it is an unusual application whose claims are allowed without any modification, the first office action will normally involve a rejection based on either formal grounds, prior art, or both. In some cases the examiner discovers and cites prior art that defeats patentability either because it anticipates the claimed invention or creates an irrebuttable case of obviousness. In this situation the applicant has little choice but to abandon the application. In those cases where the applicant is not surprised by the citation of prior art that defeats the patentability of the claimed invention, it is normal practice to rebut the examiner's rejection of the claims. The rebuttal may require amendment of the claims either to make them comply with formal requirements or to delineate more clearly the differences between the claimed invention and the prior art. Claim amendments are freely permitted but only to the extent that they find support in the specification as filed. When limitations not found in the original specification are necessary to establish patentability, they can be brought in by refiling the application. A disadvantage to this practice is that the added limitations will be effective only as of the filing date of the refiled application and may be rendered unpersuasive of patentability by having been disclosed elsewhere after the filing date of the original application but more than one year before that of the refiled application. This emphasizes the need for careful preparation of each patent application to be filed, as some mistakes are not correctable.

In those cases where rebuttal argument with or without claim amendment is not persuasive of patentability, reliance on the objective indicators of unobviousness previously discussed becomes necessary. If supported by the facts, a showing by affidavit that the claimed invention is unexpectedly superior to the prior art and has achieved extraordinary commercial success, can be quite persuasive. Any such affidavits must be very carefully prepared since they will be thoroughly scrutinized during enforcement of the patent, as occurred in the case involving 3,4-DCPA.

Another means of rebutting an examiner's rejection involves making a showing that the claimed invention was made before the publication date of a reference relied upon as defeating patentability. This practice is useful only when the application was filed no more than one year after the publication date of the reference because of the one-year grace period provided by our patent statutes. Let us say, for example, that an article is published in a British journal on January 1, 1982 describing or rendering obvious the invention claimed in an application filed on January 1, 1983. If it can be established, by some means such as reference to laboratory notebooks, that the claimed invention was made in the United States before January 1, 1982, the journal article can be removed as a reference and the United States patent application will be allowed. Even if the invention were not completed until after January 1, 1982, the reference could still be removed upon a showing that the claimed invention had been conceived before that date and diligence had been exercised in actually making the invention (reducing it to practice). Note that if the application had not been filed until January 2, 1983, this procedure could not be utilized because more than one year would have passed between the publication of the reference and the filing of the application. This procedure is not available in most countries other than the United States, since their patent statutes provide no grace period.

Interference Procedure

The above scenario involves the situation where another inventor chooses to publish the invention and thereby dedicate it to the public. A different set of rules applies when a rival inventor also seeks exclusive rights to the invention by filing a patent application. In the vast majority of countries the solution to this problem is straightforward: the first to file a patent application is entitled to the patent. The first-to-invent approach followed by the United States is unique among the industrialized countries. The remainder of this discussion will involve interference, the process used to determine which of two or more rival patent applicants was to first to invent a certain procedure. Only a small percentage of U.S. patent applications interfere with each other—that is, claim the same invention—but those that do tend to be of commercial importance. The catalytic process for the preparation of isotactic polypropylene is an example. The fact that the process was developed by multiple inventors at approximately the same time is probably accounted for by the need for such a product and the sudden availability of the necessary catalysts, rather than its having been obvious. This long-felt need accounted for the commercial success of crystalline polypropylene.

In order to prevail in an interference, the first to invent must establish that his invention date is earliest. In a typical situation, the rival inventors will have filed their applications in ignorance of the other's activity and the PTO will either initiate the interference proceeding or issue the application of the first to file (senior party) and allow the second to file (junior party) to initiate the interference. It is possible for a junior party to file an application after issuance of the senior party's patent and still provoke an interference. The junior party's application must be filed no more than one year after the issuance of the patent because of the one-year statutory grace period. In this situation the junior party must prove beyond a reasonable doubt that he was the first inventor. This heavy burden of proof, together with other consequences of filing after issuance of the rival inventor's patent, make it very difficult for the junior party to prevail. Even in the more usual situation where both parties' applications have been copending, the junior party must prove that his date of invention was before the senior party's filing date. The senior party can rely on his filing date and do nothing unless the junior party can establish an earlier date of invention (priority). A party to an interference can establish priority in one of three ways.

One means of establishing priority is to prove actual reduction to practice before the other party's actual or constructive reduction to practice. Actual reduction to practice means the making and testing of the subject matter of the invention. In the chemical arts and life sciences it is usually necessary to make and test the invention to establish its utility. However, it is possible to obtain a valid patent without an actual reduction to practice if the subject matter is such that construction and testing are not necessary to establish the invention's utility. An assembly of elements that, in combination, form a new and useful machine may well be unobvious even when the function of the machine is apparent without it having been built. When a patent application is filed describing the invention in sufficient detail to enable one skilled in the art to manufacture and use it, the invention has been constructively reduced to practice. The date of reduction to practice, either actual or constructive, can be relied on as the date of invention for purposes of determining priority.

Another way to establish priority involves a showing of constructive reduction to practice before the other party's actual or constructive reduction to practice. Basically this involves reliance on the filing of an enabling application before the other party's filing date or date of actual reduction to practice.

A party who is second to reduce the invention to practice can be awarded priority if he was the first to conceive the invention and worked diligently from the time of conception until the invention's reduction to practice. The continuum in figure 6-1, in which the abscissa represents time, is designed to illustrate this concept.

Party B was last to reduce to practice and, if this were the only

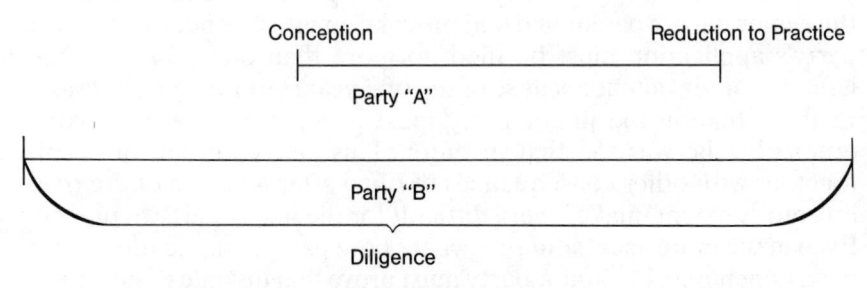

Figure 6-1. Continuum representing patent priority.

consideration, party A would be awarded priority of invention. However, party B was the first to conceive the invention; and because he was diligent in reducing it to practice, our interference rules treat the date of conception as the invention date. Under this set of circumstances party B is awarded priority. One might ask why, if party B was diligent, did party A reduce to practice in a shorter period of time? This is because reasonable diligence is the standard. While party A was more diligent (or possibly more efficient), party B was diligent enough to satisfy the reasonableness standard. It is difficult to predict what sort of activity will satisfy this requirement since these cases tend to be decided based on all of the circumstances encountered by the party trying to establish diligence. Lapses in activity while awaiting analytical results or materials that were duly ordered but delayed in transit will probably not be held to have broken the chain of diligence. Time off for illness and even a vacation have been held to be reasonable lapses in activity. However, setting aside a project to work on another even when ordered to do so by a supervisor has been consistently held to break the chain of diligence.

Proving the date of a constructive reduction to practice is not problematical because the fact that an enabling patent application was filed on a particular day can be established by the application itself. Controversy can occur on the issue of whether the application provides enablement—that is, teaches one skilled in the art how to practice the invention. Proof of actual reduction to practice and diligence, if it is a relevant factor, presents serious problems of proof to the party having the burden of coming forward with evidence. Referring back to the continuum, the conception, diligence, and reduction to practice by party B will do him no good unless he can prove them to the satisfaction of the tribunal deciding the issue of priority. This leads us to a discussion of record keeping in the laboratory.

Laboratory records can be used to establish conception of an invention. Since conception is wholly within the inventor's mind, it can be proven only through disclosure to others. In order to be effective, the conception must contain each element of the invention, including its anticipated utility if this is not apparent from the invention itself. The conception should be sufficiently detailed so that no additional invention, only construction, is necessary to reduce the invention to practice. A sketch of a mechanical device or a circuit diagram may suffice in the mechanical and electrical arts. In the case of a chemical compound, this would involve setting out a scheme for its synthesis as well as a proposed utility for the compound. A valid conception is more than just an abstract goal; it must include a means of accomplishing that goal. A conception can best be proved by its disclosure in a written document that is dated and signed by the inventor and a witness who cannot be regarded as a joint inventor. This last point is critical because joint inventors are regarded as a single inventive entity. It is axiomatic that an inventive entity cannot corroborate its own activity.

Reduction to practice (the actual carrying out of the invention) is proved by the recital of acts carried out by the inventor; these acts must be corroborated by someone who is not a joint inventor. The existence of properly kept and witnessed laboratory notebooks will not establish reduction to practice in and of themselves. It has been held that all they do is establish the existence of the notebooks as of a certain date, not that the invention was actually made as of the date of the notebook entries. Up until 1975 the corroboration of reduction to practice was very difficult and could only be established by one who could swear that he either watched the inventor carry out all the important aspects of the invention, or himself carried out the invention on behalf of or at the behest of the inventor.

The first means of corroboration under this test, which has become known as the *per se rule,* does not come into play often because research is not normally carried out before an audience. However, the second means was and remains a very important factor in cases where corroboration of reduction to practice is necessary. The technician who actually carries out the physical steps necessary to reduce the invention to practice, but is not a joint inventor, is the ideal individual to establish that the invention was completed as of a certain date because as a noninventor his testimony requires no further corroboration. Just the opposite would occur if the technician were incorrectly joined as an inventor since third party corroboration would then be required. Irrespective of the question of the patent's validity, a misjoinder of inventors can result in the loss of an interference that otherwise might have been won. The possibility of correcting a misjoinder should not be relied on when it is clear that an individual has not made an inventive contribution, for not all misjoinders are correctable. Furthermore, an attempt to remove a misjoined inventor after declaration of an interference can cast a cloud over the credibility of the party making the attempt, since it will appear as if the motivation for correcting inventorship is to "manufacture" a witness who needs no corroboration. When the other party is awarded priority, it does not simply mean that the losing party's research organization does not receive a patent. The prevailing party is placed in a position to receive a patent that can be enforced against all others, including other parties to the interference. This can be a serious consequence for failure to follow the correct procedures in keeping laboratory records and/or in determining inventorship.

The apparent rigidity of the per se rule for corroboration has been repeatedly challenged by the patent bar as causing injustice in some situations. These challenges resulted in the adoption of the rule of reason starting with a case decided in 1975. This case involved an interference the subject matter of which was a xerographic photoreceptor. The inventor had made the device himself in the privacy of his laboratory, and, while proper notebook records were kept, there was no basis for corroboration of the reduction to practice under the per se rule. However, the inventor had given the completed device to a technician to determine if it could be used to reproduce an image. The utility of the device was confirmed and testimony by the recipient of the completed device was held to be corroborative of the inventor's assertions. The tribunal found that in the absence of any evidence tending to establish that the device had been made by a third party, its transfer from one asserting inventorship to the technician added credibility to the otherwise uncorroborated reduction to practice. It should be noted that had the inventor in this case found it necessary to establish diligence in order to carry the date of invention back to the time of conception, he could not have done so because there was no corroboration until the device was completed and ready for testing.

In a later case the rule of reason test was extended to include the olfactory sense. The interference involved a chemical synthesis in which hydrogen sulfide was formed as a by-product. One of the parties was able to establish reduction to practice by reliance on properly kept notebooks whose authenticity was corroborated by a coworker who testified that there was an odor of "rotten eggs" coming from the inventor's laboratory during the period under consideration.

Under the rule-of-reason approach, proof of reduction to practice is accomplished by weaving a web of facts that will convince the tribunal beyond a high probability of fabrication that the inventor did what he claims to have done. The basic element is still well-kept laboratory notebooks, but the method of verifying their authenticity has been liberalized. The rule-of-reason approach is no panacea since there must be additional evidence that can be relied upon for corroboration. If the condensation reaction described above had been one that produced H_2O instead of H_2S, there would have been no basis for corroboration; and it is still believed that the party who can rely on the testimony of one who assembled the invention but is not a joint inventor is in the best position to establish his date of reduction to practice.

LABORATORY NOTEBOOKS

This discussion is not intended as a primer on the mechanics of keeping laboratory notebooks, but a few guidelines that are of special interest for patent purposes should be included for the sake of completeness. The notebook entry should be sufficiently clear to enable a reader who is skilled in the art to understand what was in the inventor's mind in the case of conception and exactly what was done in the case of reduction to practice. Definiteness and clarity are of course desirable attributes. Entries should be made (and witnessed) contemporaneously with what is done; delay of even one day can sometimes diminish the value of a record. In making entries, an inventor should always describe what happened, not what was hoped would happen. Only actual facts and dates should be entered and no erasures should be made. A mistake can be corrected by drawing a single line through the incorrect entry and writing in the correct version. An entry should never be altered at a later date, nor should blanks be left in the notebook lest the inference be made that other blanks were filled in at a later date. The use of ink or indelible pencil will help to rebut an assertion that alterations were made. If the researcher wishes to start on a fresh notebook page each day or for each project, any unused portion of the previous page should be X'd out before beginning on the new page. It is desirable to include sources of materials, especially chemicals, to establish that the scientist was actually using what he thought he was. The use of abbreviations should be relied upon only if the record sets out what they stand for. Finally, the notebook entries should be made as if one who is motivated to defeat the patentability of the invention will be allowed to read them. This is exactly what will happen in the case of an interference or an infringement suit involving the enforcement of any patent resulting from the research effort. Accordingly, negative notations such as "n.g." or "no R.T.P." should be avoided. Our concept of reduction to practice does not require that an invention be refined to the point of commercial feasibility, only that the sought-after result be demonstrated.

LICENSING TECHNOLOGY

The laboratory manager may be asked to provide technical expertise relating to ongoing or contemplated licensing negotiations, so some discussion of the licensing of intellectual property is appropriate. The licensing of proprietary information (patent rights or know-how) permits the licensee to use the technology licensed by the licensor. The rights granted may be exclusive or nonexclusive, worldwide or limited to a geographical area, or limited in some other manner. A license to intellectual property will typically involve business and legal considerations that are too numerous to mention here. Suffice it to say that a properly negotiated license agreement should benefit both parties by providing for a mutually satisfactory business relationship. Experienced negotiators know this and do not define winning as beating the other side but instead try to find out what the parties really want and need for a successful business relationship.

INVENTORSHIP AWARDS

It may be appropriate to close this chapter with a brief discussion of rewarding employed inventors for making inventions and discoveries. The policy on providing such rewards is normally made at a fairly high level in the organization sponsoring the research and will vary greatly from organization to organization. At the time of filing a patent application, some companies provide a cash award, which in some cases is supplemented at the time of issuance of the patent. Award plans sometimes provide for inventor compensation when a patentable idea is maintained as a trade secret for business reasons. Other systems provide for evaluation of the significance of the invention to determine the amount of the award, while some companies take the position that the employed inventor is adequately compensated by his salary and no further incentive is required. Regardless of what program is in place, the laboratory manager can reward an individual's innovation and provide incentive for additional contributions at salary review time. This sort of system may be more meaningful than the more formal approach because the laboratory manager is in the best position to judge the contribution for its qualitative and quantitative merit.

SUMMARY

This chapter began by defining proprietary information and then discussed two methods of maintaining its exclusivity (trade secret or patent protection). After a discussion of the nature of patentable subject matter, the trilogy of prerequisites for patentability (utility, novelty, and unobviousness) were explained in some detail in order to give the laboratory manager some feel for whether or not his staff has come up with a patentable invention. At this point, the chapter explored the issue of determination of inventorship and the elements of the patent application itself. A section on keeping laboratory notebooks was preceded by a detailed description of interference practice in order to emphasize the importance of proper record keeping. The chapter closed with some brief comments on licensing technology and inventorship awards.

REFERENCES

Brenner v. Manson, 380 U.S. 971, 145 USPQ 743 (Sup. Ct. 1965).
Diamond v. Charkrabarty, 100 SC 2204, 206 USPQ 193 (Sup. Ct. 1980).
Diamond v. Diehr & Lutton, 209 USPQ 1 (Sup. Ct. 1981).
Graham v. John Deer Co., 379 U.S. 956, 144 USPQ 780 (Sup. Ct. 1966).
In re Papesch, 315 F^{2d} 381, 137 USPQ 43 (CCPA 1963).
Kewanee Oil Co. v. Bicron Corp., 416 U.S. 470, 181 USPQ 673 (Sup. Ct. 1974).
Mattor v. Coolegem, 530 F^{2d} 1391, 189 USPQ 201 (CCPA, 1976).
Monsanto Co. v. Rohm & Hass Co., 312 F. Supp. 778, 164 USPQ 556 (DC EPa, 1970).
Sears, Roebuck & Co. v. Stiffel, 376, U.S. 225, 140 USPQ 524 (Sup. Ct. 1964).
Section 103 of Title 35 of the United States Code.

PART 3
GOVERNMENT–ITS IMPACT ON RESEARCH MANAGEMENT

Chapter 7
Federal Budget Cycle: Impact On Science, Technology, and Research Management

HISTORICAL PERSPECTIVE

There is little doubt that science, and the knowledge derived from it, has become a major factor in the affairs of mankind. Indeed, Western nations have come to support scientific endeavors with the expectation that, as a consequence, progress will be made toward some intended, practical goal. The scientific enterprise is supported as a mission, with goals set by governments in their own interest and in the interests of their constituents.

In the United States there is no specific, detailed, national science policy as such. Rather, the factors impinging upon the scientific enterprise are an integral part of our national policies. Science has become, in essence, a tool or extension of our national policy, which establishes scientific priorities. National policy, defined as broad public purposes, transcends and constrains the formulation of any single science policy. Thus, however broadly defined the terms and conditions of federal support for academic science, they are inevitably limited by the legislative and program restraints of the mission-oriented agencies providing support. To emphasize this, it may simply be pointed out that federal agencies such as the National Institutes of Health (NIH) and Office of Naval Research (ONR), which provide substantial levels of research support, are health and defense agencies respectively, not science agencies (Murtaugh 1973).

To examine the evolution of events that led to the ensconcement of science in the niche it currently occupies, we need hardly go back further than World War II—for it was only after this tumultuous period in our history that public appreciation for scientific endeavor began to grow. With the knowledge that closely directed, herculean research efforts yielded such advances as the atomic bomb (some would question this) and mass production of penicillin, it was no wonder that the public and the government began to view science as the architect of the future. After all, only three years after the first self-sustained, controlled nuclear chain reaction was accomplished at the University of Chicago, the Manhattan Project, at a cost of $2 billion, effectively ended the war with the detonation of two atomic bombs (Gowing 1983). Furthermore, it became apparent during the war that only through government effort was mass production of penicillin able to come to fruition. Public enchantment with science and its technologic innovations, grew, perhaps fanned by Vannevar Bush's report, *Science: The Endless Frontier*, published in 1945. The 1950s saw the rapid expansion of budgets for activities that were deemed appropriate societal goals (peaceful development of atomic energy improved, health care, and so on). New impetus was also provided during this period to American academic science, particularly those areas related to space and defense (physics and mathematics) after the successful launch of Sputnik I. Between 1956 and 1967, federal expenditures for basic research rose from around $200 million to about $2 billion—an average annual growth rate of about 21 percent (Haskins 1973). By the late 1960s, however, when the United States was contending with the burden of costly international commitments and the depletion of resources resulting from the Vietnam war, the "golden era" of research spending came to an end. The growth rate of expenditures for basic research dropped precipitously from 21 percent per annum to around 7 percent. The impact upon the scientific enterprise and community was both dramatic and traumatic. The decline in government support occurred during a period of increasing social unrest and shifting popular attitudes. The public was, in effect, asking if our national priorities reflected our national needs. The situation was compounded by concerns over damage to the environment and the quality of life, as well as our dependence upon foreign energy sources and the impact of this dependence on our national economy in the 1970s. After all, a sound, healthy economy is essential to the scientific enterprise (the number of scientific papers published bears an almost constant ratio to the Gross National Product among advanced nations) (Brooks 1973).

The public uneasiness with science and its purposes came at a time when science and technology had brought man to a pinnacle of achievement—the first manned extraterrestrial exploratory mission. What for some had seemed to be a straightforward, natural relationship between science, the technology it spawns, and meritorious social purposes now seemed to go awry. It had been reasoned that as the scientific base of knowledge expanded, technological innovation would naturally follow (Shannon 1973) and in turn affect social change— exemplified by an enhanced quality of life for our citizens. The consequences would be reflected in our social value system. Ultimately this meant that society would see the benefit in further expansion of the base of scientific knowledge and thus fuel (with funding) another cycle leading to further social changes and improvement. With a little fine-tuning, this cycle could operate in homeostatic perpetuity. However, the social unrest in the 1960s and 1970s acted as a negative feedback inhibitor of the science/social change cycle.

Were the criticisms of science and technology justified? Some would argue that technology is neutral in itself, and that the criticism of the scientific community should have been directed to the broader social system per se, for it is here that responsibility lies in regard to the manner in which science and technology are utilized. Another view is expressed by Chargaff (1983), who states, "[W]hat is being cooked may not be what people think they are cooking. During the entire history of science, scientists have always underestimated the consequences of what they are doing."

Whatever the merits of the arguments, one point is obvious: if science is to play a role in the purposes and goals of national policy, then the scientist and those who manage this endeavor must actively participate in the policy-making process and in integrating the base of scientific knowledge into the machinery of the social structure. The scientist must exhibit more visionary qualities with respect to the ramifications and consequences of how newly obtained knowledge is used. How the scientific community meets this responsibility will have a direct bearing on the way policy decisions are derived. Through necessity, scientists must learn to speak both the language of science and that of public policy (Kapitsa 1966) if we are to pursue our profession and maintain our scientific organizations. If science is an expression of public policy, then knowledge of the mechanisms by which the scientific enterprise is sustained is essential.

In the mid-1980s, the federal government funded about 45 percent of the total U.S. expenditures for R & D—making it the largest single provider of research monies (Duga and Fisher 1984). About 30 percent of industrial and 75 percent of academic R & D was federally funded. It is apparent that knowledge of the federal budgetary process is imperative if a research laboratory, whether small or large, academic or proprietary, with basic or applied orientation, is to compete favorably for funds. This chapter seeks to outline the context within which the scientist and/or laboratory manager must operate if the language of "science and public policy" is to be spoken and where, in the budget process, this language may be most effectively expressed. The budgetary constraints and "windows of opportunity" that are so important for those seeking grants and/or contracts are discussed in subsequent chapters.

STEPS OF THE BUDGET CYCLE

Shapley (1976) identifies 123 steps, within three major phases, that comprise the federal budget cycle: budget formulation, congressional review, and budget execution. The major budget phases, and their respective time frames are represented diagrammatically in figure 7-1. The formulation phase is divided into three steps. The first of these involves *preliminary projections* of the budget for the fiscal year (FY) under consideration, let us say FY 198X. Based upon the requirements of the Congressional Budget Act of 1974, this step would begin about six years prior to actual budget execution. Thus, the comple-

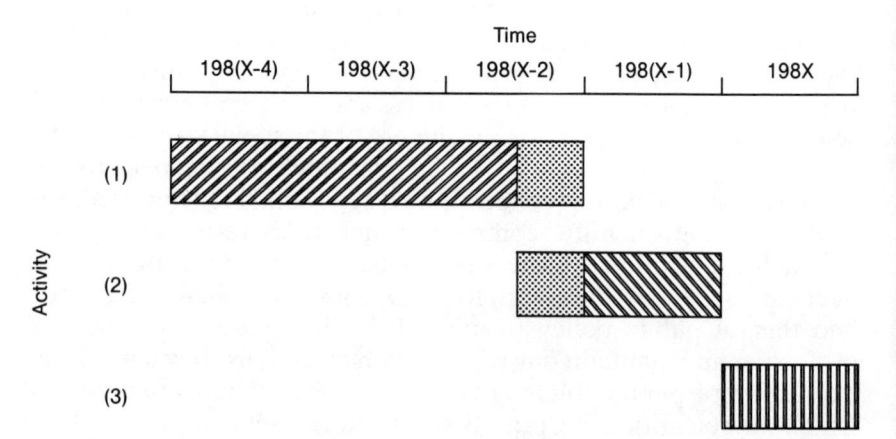

Figure 7-1. Idealized time frame of major phases of a five-year federal budget cycle. (1) Budget formulation; (2) Congressional review; (3) Budget execution. Stippled areas represent overlap of congressional review of 198X budget, with OMB and agency budget formulation phase. Overlap occurs when Congress begins review of the five-year current services projects in 198(X−2).

tion of a single budget cycle from the preliminary projections to actual expenditure of the authorized funds now spans about seven years!

The earliest projections for a particular area of funding begin with review of the FY 198X budget estimates included in the five-year budget projections submitted to the appropriate agency and to the Office of Management and Budget (OMB) with the FY 198(X−4) budget. The agency identifies programs, major projects, and their content and respective funding levels, often with advisory committee recommendations, and submits this information to OMB. The latter then prepares projections for the president, based upon the assumption that current services or programs will be continued without change. These projections are then modified to reflect the impact of the president's FY 198(X−4,3,2,1) budget decisions upon the FY 198X budget. OMB then develops specific recommendations concerning major policy items and funding levels for the FY 198X budget upon which the president will make final decisions. One must recognize that there is no R & D budget per se. The R & D enterprise is funded through a vast number of federal agencies and consequently is subject to the continually changing perceptions of Congress and the executive branch as to what comprises our nation's needs and goals, and what resources are available to achieve those ends. The formulation of the FY 198X budget to this stage will have consumed approximately three years.

The second step, the *planning and development* phase, begins when the president submits his authorization requests for FY 198X, in compliance with the requirements of Section 607 of the Congressional Budget Act, with his FY 198(X−1) authorization bill. This step transpires in the early part of 198(X−2). Within the same time frame, OMB transmits official allowance letters to agencies regarding the FY 198(X−1) budget. These letters may also provide "guidance" with respect to the FY 198X budget.

Upon receipt of the president's authorization request for FY 198X, Congress begins its initial examination in House and Senate hearings. It is quite likely that the FY 198X budget plans will be affected by House and Senate floor actions when the FY 198(X−1) authorization and appropriation bills are debated.

By spring or early summer of 198(X−2), the agency will reevaluate its plans, priorities, and estimates and resubmit them to OMB. The *planning and development* stage is culminated when OMB, in accord with the president's budget decisions, develops the proposed general budget policies, specific guidelines, and dollar targets for the respec-

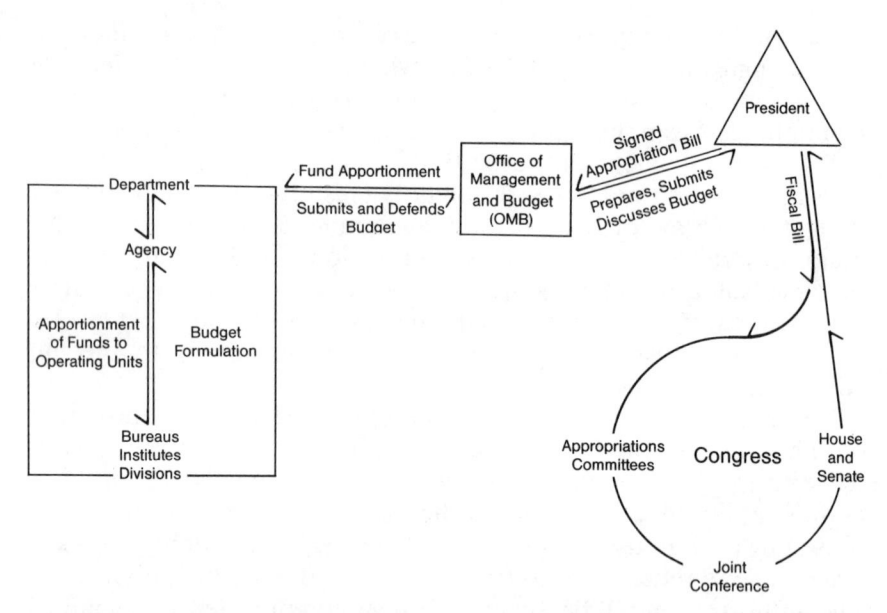

Figure 7-2. Abbreviated federal budget cycle schema. Initial budget development begins at department level, proceeds through the agency and institute to OMB, the president, and Congress. The appropriation bill is signed by the president and submitted to OMB for apportionment to the respective institute.

(such as a notice of grant award). A schema for the 198(X−1) and 198X years of the budget process is represented in figure 7-2.

The budget execution phase does not end with the award of grants, contracts, and agreements. Internal management control must be maintained over the rate and manner in which awarded funds are being utilized. The expenditures, or "outlays," are made to the awardee in accord with the terms of the obligation, usually on a quarterly basis or after the work has been performed. Reviews—both programmatic and fiscal (audits)—are made periodically and may lead to changes, redirection, or termination of funded programs. A condensed version of the steps in the budget cycle is provided in table 7-1.

IMPOUNDMENT, CONTINUING RESOLUTIONS, AND OTHER EXIGENCIES

As one must now realize, the gauntlet through which a budget must pass, in its normal cycle of genesis, is fraught with a multitude of

Table 7-1. Abbreviated Budget Cycle

PHASE I. BUDGET FORMULATION

Planning and Development Stage

198(X-5)   Preliminary projections are submitted to Agency and OMB with FY 198(X-4) budget. Agency identifies major projects and funding levels, submits plan to OMB. OMB prepares projections based upon impact of FY 198(X-4,3,2,1) budget decisions upon FY 198X budget; resubmits to President.

198(X-2)   President submits authorization request for 198X budget to Congress with 198(X-1) authorization bill. FY 198X budget plans may be altered by House and Senate actions on FY 198(X-1) authorization and appropriation bills.

Agency reevaluates its plans and budget estimates and resubmits to OMB. OMB, with President's budget decisions, develops the proposed general budget policies and dollar targets for the respective agencies.

Budget Preparation and Review Stage

198(X-2)   OMB policy and dollar targets are resubmitted to agency, followed by agency, OMB, and presidential review, and adjustments to reflect congressional action on FY 198(X-1) budget. President's budget document is published and agency's justification to Congress submitted.

198(X-1)   President's FY 198X budget and agency's FY 198X authorization bills are submitted to Congress.

PHASE II. CONGRESSIONAL REVIEW

Budget, Authorization, and Appropriation Stages

198(X-2)   These stages run concurrently and begin in 198(X-2), when Joint Economic Committee completes review of five-year current services projects. Reports are given to respective House and Senate Budget Committees. Budget committees prepare FY 198X "concurrent resolution" and submit it to full legislative body. Differences in House and Senate resolutions are resolved in conference.

Legislative committees take 198(X-1) authorization bill under consideration. Upon passage by both houses and presidential signature, bill becomes law.

198(X-1)   House and Senate Appropriations Committees begin hearing on 198X budget estimates. Differences are resolved prior to passage of appropriations bill.

PHASE III. BUDGET EXECUTION

198X   Agencies develop operating budget plan based upon appropriations bill. Request apportionment of appropriations by OMB. OMB reviews apportionment requests and agency is allowed to allocate available funds to operating subdivisions. Commitments are approved and funds obligated.

awesome impediments. At any number of steps, "rachets" may be applied to tighten the budget. But even when an appropriations bill is passed by Congress and signed by the president, a serious effort may be made to alter the nature and magnitude of those appropriations in the budget execution phase. Should the president judge that the appropriated funds are not required for their intended purpose or that they are not in concurrence with his fiscal policy, he may withhold those funds from apportionment. If this occurs, Congress may agree and comply with the president's wishes by passing a "recission bill." However, if the recission bill is not passed by Congress within forty-five days, the funds automatically become available for obligation.

Likewise, the president, OMB, or the agency head may "defer" the use of congressionally authorized funds. This may occur, however, only if neither house of Congress passes a special "impoundment resolution" to disapprove the deferral. If one of the houses passes such a resolution, the funds become available for obligation.

A new twist in fund deferral, "forward funding," has recently been introduced by OMB. In this case, the funds would be apportioned to the respective agency. Let's use the National Cancer Institute as an example. The institute would be obligated to select a small percentage of the top, competitively ranked projects and apply the apportioned funds (FY 198X) to "complete" or "forward funding" of those projects, perhaps for the next three years. Whereas this maneuver does not, at least upon first examination, circumvent the intent of congressionally approved appropriations, it could have pronounced impact upon the scientific enterprise. First, it has the immediate effect of reducing the level of funds available for new projects, and second, it introduces an added degree of instability to the continuity of funding for those projects not selected for "forward funding." The continuation of projects funded for only one year, which would represent the bulk of those approved and funded, would be subject to the exigencies of FY 198(X+1,2) budgets. Therefore, the likelihood of such a project being continued beyond the immediate year of support is highly questionable. As scientists and R & D managers are aware, the "payback" time for most scientific endeavors is considerably longer than one year. Thus, the end result of such action would not only have a profound effect upon the scientific infrastructure, but would entail fiscal waste because projects funded for only one year would rarely reach fruition.

One further impact of budget contingencies upon the scientific enterprise is not so much that imposed by any particular fiscal policy, but rather by the lack of one. If the regular FY 198X appropriation bill is not passed before the beginning of the fiscal year (now October 1), *continuing resolutions* must be reported, agreed to, and passed by both houses in order to make interim appropriations available. These continuing resolutions are passed for a specified period, usually thirty to ninety days, and new resolutions must be passed until the appropriation bills are finally approved. Recently, we have seen continuing resolutions lapse, and with no funding authority, agencies have had to furlough employees.

It is against this backdrop, a milieu of fiscal events and uncertainties, that the scientist and R & D manager must operate. To be successful in competition for limited fiscal resources and in the effective conduct of funded projects, it behooves one to be aware of the vagaries of the budget process and of opportunities—usually early on in the cycle—for the scientific community to influence its course.

SUMMARY

Since the late 1940s, the fortunes of U.S. science have been intimately intertwined with the national policy-making process and the fabric of our social structure. Science has found both utility and sustenance in the purposes of national policy and the resources allocated to achieve those goals. The mid-1980s manifest this relationship: federal monies funded about one-half of all R & D conducted in the United States and 30 and 75 percent of industrial and academic research, respectively. This relationship between scientific enterprise and national policy is likely to continue—although easily influenced by the public perception of the scientific contribution to the attainment of worthy goals (best exemplified by tangible evidence of an enhanced "quality of life") and the condition of the national economy. More and more, the scientist and those who manage the scientific endeavor, if they are to continue to ply their profession, must "match" their particular field to established goals as set forth in national policy. Thus the scientist and/or manager must be familiar with the federal budget process— must know how it operates, how it may work to their benefit, and how and at what point it may be affected to their input. Some 123 steps in the federal budget cycle have been identified, making up three major phases. Points at which the budget process may be most easily influenced, either through legislative pressure and/or advisory group recommendations, are specified. Finally, knowledge of the federal budget process sets the backdrop against which those seeking federal monies, either through grants and/or contracts, must operate.

REFERENCES

Brooks, H. 1973. "The Physical Sciences: Bellwether of Science Policy." In *Science and the Evolution of Public Policy*, ed. J. A. Shannon, 105-34. New York: Rockefeller University Press.
Bush, V. 1945. *Science: The Endless Frontier.* Washington, D.C.: U.S. Government Printing Office.
Chargaff, E. 1983. "Philosophical Direction and Social Constraint." In *Priorities in Research*, ed. Sir John Kendraw and J. H. Shelley, 56-57. Amsterdam: Excerpta Medica.
Duga, J. J., and W. Fisher. 1984. *Probable Levels of R & D Expenditures in 1985: Forecast and Analysis*, 1-20. Columbus, Ohio: Battelle.
Gowing, M. 1983. "Historical Precedents in Scientific Research." In *Priorities in Research*, ed. Sir John Kendraw and J. H. Shelley, 3-10. Amsterdam: Excerpta Medica.
Haskins, C. P. 1973. "Science and Social Purpose." In *Science and the Evolution of Public Policy*, ed. J. A. Shannon, 1-18. New York: Rockefeller University Press.
Kapitsa, P. 1964. "The Future Problems of Science." In *The Science of Science*, ed. M. Goldsmith and A. MacKay, 126. New York: Penguin Books.
Murtaugh, J. S. 1973. "The Federal Support of Science: Biomedical Sciences." In *Science and the Evolution of Public Policy*, ed. J. A. Shannon, 157-87, New York: Rockefeller University Press.
Shannon, J. A. 1973. "Introduction." In *Science and the Evolution of Public Policy*, ed. J. A. Shannon, vii-xvi. New York: Rockefeller University Press.
Shapley, W. H. 1976. *Research and Development in the Federal Budget*, 1-99. Washington, D.C.: American Association for the Advancement of Science.

Chapter 8
Grants and Grant Management

WHAT IS A GRANT?

Individuals involved in the conduct or management of science and technology—scientists, engineers, or professional managers—will almost certainly at some time in their careers become involved with grant programs and their management. Managers in the business community are more often familiar with contractual agreements, and only recently have grant programs become a viable option for proprietary organizations that seek research support. This is particularly true in the pharmaceutical and high-technology biomedical industries. Although there are thousands of private foundations, trusts, corporations, and individuals that employ the "grant" as a mechanism for rewarding past activities and encouraging future ones, the federal government is by far the richest source of grant funds (White 1975). Emphasis is therefore placed here on the federal grant mechanism, although many of the properties and conditions described apply to private grant programs as well.

There are a number of distinctions between the more conventional contract mechanisms of funding sought by industry and the traditional grant programs more frequently pursued by nonprofit organizations (Teague 1981). Therefore it is imperative to understand the differences in these two funding processes if the applicant is to compete successfully. Indeed, a comparison of contracts and grants may best serve to define the latter. Contracts will be considered in detail in the following chapter.

Grant programs differ from contracts in origin of specifications, being created almost always in response to congressional legislation. Broad goals and priorities are established by the agency or federal administrative unit responsible for implementing the provisions of the legislative act. The National Cancer Institute's "War on Cancer" is an example of one such legislatively established goal. These general goals and priorities are described in the *Federal Register* and summarized in the *Catalog of Federal Domestic Assistance*. The latter identifies the type of grant programs established, eligibility requirements, restrictions, application procedures, and program policies and regulations governing the programs. Program announcements, such as the *NIH Guide for Grants and Contracts*, serve to inform potential applicants of the specific needs and approaches that relate to agency goals. The solicitation document for the grant is simply an application kit with guidelines for proposal preparation.

As mentioned, only recently has the grant become a potential source of research funds for proprietary organizations. The reason for this, aside from previous legislative restrictions, lies in other important differences between grants and contracts. All grants are awarded on a cost-reimbursement basis and the recipient receives no fee. Contracts, on the other hand, typically provide a fee in addition to cost reimbursement. Thus the nature of the grant award has made it of limited interest to the private business sector. In addition, because the contract solicits a service or product on a one-time basis, it is more closely monitored by the granting agency in regard to performance, with more stringent deadlines and requirements for reports, specifications, and financial audits. The grant, by comparison, undergoes little monitoring, operates under minimal fiscal control, and thus has greater appeal to the academician. The distinguishing characteristics of the federal grant are listed in table 8-1.

The grant then is a mechanism whereby an agency may accomplish

Table 8-1. Distinguishing Characteristics of the Federal Grant

1. The basis for a grant is almost always established by legislative act based upon current or perceived need.
2. Grant programs are announced in the *Federal Register*.
3. Grants are normally awarded to nonprofit organizations although no longer legislatively restricted to this type of organization.
4. Multiple grant awards are made in support of a single agency goal.
5. Most grant programs are cyclical with application receipt dates occurring more than once each year.
6. Grants are awarded on a cost-reimbursement basis.
7. The awarding agency exhibits flexible financial control and minimal performance monitoring.

its broad program goals through the efforts and expertise of the scientific community that resides outside the formal framework of the federal hierarchy. Specifically, it is an award of financial assistance provided by legislative action to eligible recipients whose proposed activity or program has successfully undergone review and approval.

TYPES OF GRANTS

Grants may be awarded in support of a wide variety of activities, including a range of social programs and community services. *Formula* grants are those that provide funds to specified grantees on the basis of a specific formula, frequently demographic data, prescribed in legislation or by federal regulation. These grants, which are often awarded on a mandatory basis, are usually made to states, counties, or municipalities.

Our discussion will be limited to grants classified as *discretionary* or *project* grants, which are the types most likely to be sought by researchers and research organizations. A discretionary grant is made in support of an individual project in accordance with the legislative act that authorizes the agency to exercise its judgment in selecting the project, grantee, and the amount of the award. Among the discretionary grants, three major categories of support can be discerned—those provided to institutions for facilities support, training, and research. Several forms of *biomedical research support* grants, related to the first two categories, are listed and briefly described in table 8-2. The remainder of the discussion in this section is devoted to grants made in direct support of research activities.

Grants developed expressly for support of research activities may be awarded either directly to institutions or on behalf of an individual affiliated with the institution. *Consortium* grants are made to the former in support of a research project in which the program is being carried out through a cooperative agreement between the grantee institution and one or more participating institutions. *Resource* grants are special-purpose awards made to institutions in support of research resources. These resources may include computer centers, clinical research centers, or other facilities that are available to all qualified investigators irrespective of their scientific disciplines or special area of research activity.

Several kinds of research grants can be awarded to an institution on behalf of an individual investigator. These grants are made in support of an investigation aimed at the discovery and interpretation of facts,

Table 8-2. Special Biomedical Research Support Grants

Category	Basis for Award
Facilities assistance grants	
Planning	Supports planning, developing, designing, and establishing the means for performing research, delivery of health services, or accomplishing other approved objectives
Construction	Provides for acquisition, remodeling, expansion, or leasing of existing facilities, construction of new facilities, and for initial equipping of such facilities
Education/training grants	
Capitation	Provides for the maintenance and improvement of the institution's educational programs in health professions, nursing, and allied health areas
Training	Supports the training of students, personnel, or prospective employees in research or in practices and techniques that relate to the delivery of health services
Fellowship	Made on behalf of an individual in support of specific training to enhance that person's level of competence
Conference	Awarded in support of costs of meetings, such as workshops and symposia, in which "state-of-the science" information is disseminated
Demonstration	Used to establish or demonstrate the feasibility of a theory or approach. Generally of limited duration

revision of accepted theories in light of new facts, or the application of new or revised theories. Most common is the *research project grant*, awarded for a discrete research project in an area of the investigator's interest and expertise and focusing upon a single scientific problem or objective. *Program project* grants, *center* grants, and *research career development awards* are also made to institutions on behalf of individual investigators, with the express purpose of providing a specialized kind of support to research activities. Program project grants support long term programs having a specific major objective. These projects are often multidisciplinary and involve the efforts of groups of investigators. Center grants are awarded on behalf of a program director and a group of collaborating investigators. They, too, support long-term multidisciplinary research programs but may also be used to support demonstrations, education, and other related nonresearch components. A major difference between program projects and center grants is that the latter are usually developed in response to specific announcements of the programmatic needs of the responsible agency. In regard to specification development, the center grant resembles a "hybrid" between grants and contracts. The third type, research career development awards, provides a mechanism to foster the early development of scientists who demonstrate outstanding potential for research careers in the health-related sciences. These awards provide all or part of the recipient's salary.

Within each of the broad categories of grants described, several types have been delineated. For example, at the National Institutes of Health (NIH) a "type 1" refers to a new competing application; "type 2," to a competing renewal; "type 5," to a noncompeting continuation grant; and so forth. Since the NIH is the major provider of federal funds for health research in the United States, we will describe, in general terms, the grant mechanism employed by that agency to achieve its program goals. One must be mindful that, while other federal granting agencies may provide support to research and research-related areas via similar mechanisms, their specific goals will differ. Therefore specifications, eligibility, and other particulars should be carefully investigated prior to application. This brings us to the next major focus.

THE FIRST STEPS IN APPLYING FOR A GRANT

The period prior to formal application to a funding source involves considerable planning and groundwork. The outcome of your quest for research support will very often depend upon how thorough a job you have done in this phase of proposal planning. Serious consideration must be given three major questions (Murry and Biles 1980). First, are you certain that what you are proposing to do is innovative and responsive to some pressing issue or problem? Second, do you and the other individuals associated with your project possess the background, experience, and expertise to carry the project through to a successful conclusion? Third, is the institution with which you are affiliated both capable of and willing to provide the necessary support to the project?

In order to answer the first question and eventually make the correct "match" to your project, you must first identify potential sources of support. There are over 25,000 private foundations and 1,200 federally funded programs to be evaluated as potential funding sources for your project. Most of the goals and objectives of the federal agencies can readily be identified from program guidelines published in the *Federal Register* or listed in the *Catalog of Federal Domestic Assistance*. These publications, which can be obtained from the U.S. Government Printing Office, contain other pertinent information such as eligibility requirements, levels of available funding, specifications, application procedures, and deadlines, as well as review criteria.

The task of narrowing the list of potential sources from private foundations is a much more formidable one. However, information concerning foundations and their objectives can be obtained from the Foundation Center. Source materials, such as the past and current granting activities of each foundation, are maintained by the Foundation Center in national repositories located in Washington, D.C.; New York; and Chicago. Regional collections contain information directly relevant to your specific geographic location. Resource publications, *The Foundation Directory* (published biannually) and *The Foundation Grants Index* (published annually), provide specific information such as lists of grants awarded, size of awards, and subject fields in which the awards were made. These publications are available in the regional Foundation Center collections and in many academic and public libraries. Your institution's office of development, or its equivalent, may also be able to provide you with this type of assistance.

It is at this point that some explanation is in order regarding the responsibilities and roles that the professional laboratory manager and principal investigator share in the grant application process. We have assumed that in many laboratories a principal investigator will also wear the hat of manager. Consequently, we will lead the reader through the entire process of applying for a grant, from conception to management. Whereas the professional manager may not be expected to have the scientific and technical expertise required to assure scientific validity of a research proposal, he must certainly be knowledgeable concerning all other aspects of the grant application, review, and management processes. Thus, while our comments are at times directed to the principal investigator, they are equally applicable to the professional manager who must administer research programs involving grants.

Once you have determined which foundations and agencies support goals and needs compatible with your project, specific guidelines and application materials should be obtained. These guidelines should be carefully studied with respect to applicant eligibility, deadlines, application and review procedures, and suitability of your organization in meeting the requirements of the prospective funding source.

You may find that many private foundations make awards only to nonprofit institutions and further refinement of the list of prospective funding sources may be warranted. Further, in the process of requesting guidelines and application information, you may also find that many private foundations do not enthusiastically respond to such requests. Almost certainly they will not act favorably to unsolicited applications that seemingly appear spontaneously in the mail. It will usually be necessary to initiate some kind of preproposal contact with officers or administrators of the respective foundation.

The *letter of intent* should be the first approach to a private foundation. This letter should be brief but include the following information:

- The reason for writing to the particular foundation (explain how their goals and your project's objectives coincide and complement each other)
- A short description of your research plan
- Your credentials and those of your collaborators (a resume will suffice)
- An estimate of the total funding required for the project

This letter should be addressed to the appropriate program officer or administrator, preferably by name. Names of foundation officers can usually be obtained by simply placing a telephone call.

In most cases, the letter of intent, in which the research plan is briefly outlined, will suffice in answering questions the potential granting organization may initially have. If after three to four weeks you have received no response to your letter, however, it would be prudent to follow up with a telephone call. Be forewarned that personal communication may lead to further and more penetrating questions concerning your project, your credentials, or the credibility of your sponsoring institution. Be prepared and have your case "in concrete" before making personal contact. Keep in mind that direct discussions with foundation officers can be extremely valuable in helping you adjust the content, approach, and fiscal constraints of your application so that it may satisfy the needs of both the granting organization and your project.

Although federal agencies usually provide application kits with very thorough instructions and guidelines for proposal preparation, some program directors encourage preapplication communication and may even require submission of a *preproposal*. This is especially true of extramural programs that are administered through intramural laboratories. The program directors are necessarily concerned with the applicability of your proposed activity to their laboratory goals and how much it is going to cost. The preproposal usually takes the form of a condensed application and includes the objectives and rationale of the proposed project, a general description of how the objectives are to be achieved, significance of the work, and an accurate projection of costs. The proposed budget in the preproposal should be itemized by categories—salaries, supplies, equipment, and so on.

By now you have received responses to your letters-of-intent and preproposals. You are about ready to prepare the formal application. Before proceeding, however, the instructions and guidelines should once more be carefully examined for (1) deadline dates, (2) proposal evaluation criteria, and (3) how your proposal must be organized to comply with those guidelines. In regard to deadlines, allow yourself adequate time to prepare the finished, polished proposal. Remember to allow sufficient time for any internal organizational review such as the Human Studies or Animal Studies Committees, or any others that must approve your application. Also remember that it takes time to process your application through the appropriate institutional procedures and personnel. Be certain whether the deadlines stipulate "post-marked" or receipt dates.

Knowledge of evaluation criteria used in judging your proposal is important if you are to make the proposal responsive to granting the institution. If you are to communicate your ideas effectively, you must have some knowledge of who reads your grant application and the criteria upon which it is judged. The NIH provides a guide for its study section reviewers that you will find useful. Questions posed include the following: Are the aims logical? Is the approach valid and adequate? Are the procedures feasible? Will the research produce new data and concepts or confirm existing hypotheses? What is the significance and pertinence of the proposed study with regard to the state of the field and the importance of its aims? Other criteria include the investigator's competence and background, the adequacy of institutional resources and the environment in supporting the conduct of the study, and whether the budget is realistic and adequately justified. The "Proposal Checklist and Evaluation Form," designed for use in the Grantsmanship Center training programs, is also useful in assisting the applicant to prepare an effective proposal (Kiritz 1979). Finally, in the formal application, it is imperative that you follow the guidelines in regard to organization, format, and length. The critical path in preparation and focus of a formal research proposal is shown in figure 8-1.

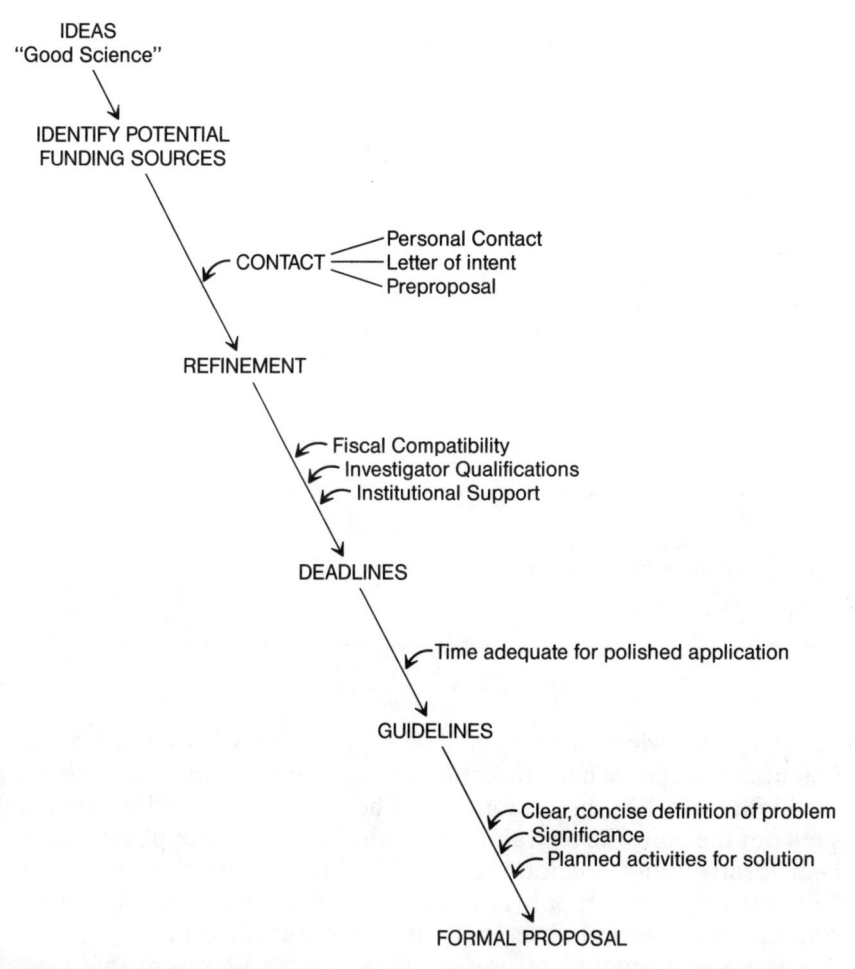

```
IDEAS
"Good Science"
    ↓
IDENTIFY POTENTIAL
FUNDING SOURCES
    ↓
         CONTACT ── Personal Contact
                 ── Letter of intent
                 ── Preproposal
    ↓
REFINEMENT
    ↓
         ── Fiscal Compatibility
         ── Investigator Qualifications
         ── Institutional Support
    ↓
DEADLINES
    ↓
         ── Time adequate for polished application
    ↓
GUIDELINES
    ↓
         ── Clear, concise definition of problem
         ── Significance
         ── Planned activities for solution
    ↓
FORMAL PROPOSAL
```

Figure 8-1. Critical path in the preparation and focus of a formal research proposal. The diagram represents steps in the genesis of a research proposal, from conception to preparation of the polished, formal application.

FORMAL PROPOSAL PREPARATION

The focus of the formal proposal will be greatly influenced by the thoroughness of the job you have done in the preapplication stage. From a list of potential funding sources identified with expressed goals, you have focused only upon those whose needs and scope are compatible with the objectives of your laboratory. Further refinement,

Table 8-3. Application Formats

I. *NIH*
   Research plan
   A. Specific aims
   B. Significance
   C. Progress report/preliminary studies
   D. Methods
   E. Human subjects
   F. Laboratory animals
   G. Consultants
   H. Consortium or collaborative agreements
   I. Literature cited
II. *General*
   Scientific summary
   Layman's summary
   A. Introduction
      1. Status of current work in area
      2. Rationale
      3. Specific aims
   B. Methods of procedure
   C. Significance of research
   D. Literature references

---

based upon review criteria, has been achieved by selecting those potential sources where investigator competence and institutional credibility can best be established. The instructions and guidelines spell out the editorial restraints. Thus the focus of the application has been pretty much decided for the investigator by the process of "natural selection." It only remains for the investigator to express his own creativeness and ideas in the most favorable manner.

There are a number of useful articles in the literature that offer sound advice on how to write a successful research proposal (Eaves 1978, Jagger 1980, Kiritz 1980, Abeles 1982). How one approaches such an undertaking will depend a great deal upon the foundation or agency requirements and the individual's level of skill in writing scientific literature. A brief description of a "typical" proposal format and what should be included in each section follows. (Two types of application formats are presented in table 8-3.)

Research Plan

The research plan is usually organized into four or five sections and addresses the four major questions commonly required in scientific papers, although not necessarily in the same order. For example, in the NIH format the *specific aims* section is simply a concise verbal description of the work you are proposing to do. The description should realistically reflect what you intend to accomplish and the hypothesis to be tested. It normally will consist of about three or four specific points. Remember that research grants are awarded to projects that focus upon a single scientific problem or objective and the specific aims should reflect this central theme. Do not describe specific experiments and do not list an unrealistic number of aims that could not possibly be accomplished during the grant period. Together with the summary or abstract, the specific aims section will provide the first impression of your project that the reviewer gets. It should be a straightforward, clear, and concise description of what you plan to do.

The *significance* section of the NIH format is a multifunction category that addresses several points. A statement of the problem the project proposes to address and a review of the relevant literature are required. The section should include concise, analytical background information that critically evaluates existing knowledge, identifying and emphasizing gaps in the body of knowledge that your proposal intends to fill. Finally, the importance (significance) of the proposed research should be succinctly stated and the specific aims of the project related to longer-term objectives.

The *progress report/preliminary studies* section requires that the applicant summarize published or unpublished results of work conducted toward the achievement of stated specific aims. This section allows the investigator of an ongoing project to demonstrate to the reviewers how productive he has been in accomplishing the grant's objective, to summarize the importance of his findings, and to discuss any changes in direction of the project from the previous reporting period.

If the project is a new one, the preliminary studies section should lean heavily upon your previous experience in the area of research being proposed. It is likely that what you are proposing to do will simply be a redirection of your previous investigations or of the knowledge gained from them. Keep in mind that most granting agencies place more confidence in researchers with a proven track record and in projects that have a promising future. This section should be used to demonstrate to the agency that, based upon past accomplishments, you are particularly suited to carrying the proposed studies through to fruition.

The *methods* section is a detailed description of the experimental design and the methods or procedures employed to accomplish the project's specific aims. The experimental design should adequately describe the protocols and the sequence in which each is to be used in the investigation. Needless to say, the protocol sequence should follow a logical order.

The procedures should be described in enough detail to allow the reviewers to decide whether the methods are scientifically sound and suited to your expressed purposes. Means by which the data are to be analyzed (statistical test methods) and how you will interpret the results should also be included. If your proposal utilizes new methodologies, you will need to justify their use by describing their advantages over existing methods. If there are obvious difficulties or limitations to the proposed methods, then you should cite alternative methods and explain why those you have chosen are most appropriate under these specific experimental circumstances. If any of the procedures or materials used in your studies present a hazardous or dangerous situation to personnel, NIH requires that you adequately describe the precautions to be taken. Although you should also strive to be concise in preparing the proposal, the methods section should be described in enough detail to allow the reviewers to judge the feasibility of your approach and the adequacy of your methods in achieving the stated specific aims.

Beyond the four major questions that every proposal must address, whether it is prepared for a federal agency or private foundation, information concerning the use of human subjects, animals, consultants, and collaborative agreements must also be provided. In some cases, the requirements for this information have been previously fulfilled by assurances of compliance from the sponsoring institution. If your project involves human subjects and derived materials or data, NIH requires the following information:

1. Identification of the sources of potential subjects, materials, or data
2. Characteristics of the subject population, the rationale for its use, and why it must be used
3. How subjects are to be recruited and consent procedures to be followed
4. A description of potential risks and their likelihood
5. Procedures for protecting against or minimizing any potential risks and an assessment of their effectiveness
6. Confidentiality safeguards
7. Potential benefits to be gained by the subjects and society as a result of their participation in the study
8. A risk-to-benefit assessment

Verification of the institutional review board's review and approval of your protocol must be submitted. It is judicious to provide a copy of the informed consent document as an addendum to your application. Further details concerning studies involving human subjects are provided in appendix A.

When laboratory animals are to be used in investigations, the researcher must identify the species, strain, age, and number involved. The rationale for their use and a description of animal care procedures must be provided, particularly the procedures required to avoid unnecessary discomfort, pain, or injury to the animals. If the application is to be submitted to NIH, the grantee must adhere to the principles of animal care and use as provided in the *Guide for the Care and Use of Laboratory Animals*. In addition, the grantee institution must file an "Animal Welfare Assurance Statement" with the Office for Protection from Research Risks (OPRR), NIH. This statement commits the grantee to comply with stringent conditions with respect to the care and use of laboratory animals in the research project. Details of requirements for animal experimentation may be found in appendix B.

Consultant arrangements need to be confirmed in writing and a letter from each consultant, describing his role in the proposed research, attached to the application. For consortium or collaborative agreements, a detailed explanation of the programmatic, fiscal, and administrative arrangements between cooperating institutions must be provided, along with confirming letters of the written arrangements.

Two other topics are very important to the treatment your grant will get at the hands of the funding agency—the *title* and the *summary*. The title needs to describe the project objective accurately. Remember that, based upon the proposed title, the initial decision may be made as to appropriateness of your proposal to foundation or agency goals and—if it is a NIH application—the institute to which your grant will be assigned.

Because of the volume of applications to government agencies, the summary or abstract may be all that is read in the initial screening, and it thus assumes great importance. At which point in the sequence of formal application preparation the summary is written is a subject of debate. John Jagger (1980) suggests that many may wish to prepare the summary first because it is useful in providing a preliminary statement of the overall problem from which the details of the proposal can be developed. Norton Kirtz (1979), on the other hand, recommends that the summary not be prepared until you have completed the proposal and can then get a better grasp or overview. There is agreement, however, upon what a good summary should contain. The summary frames the proposal and presents a statement of the overall problem, identifies the objectives of the project in addressing the problem, and provides a general statement of the methodology to be used in achieving the objectives. In addition, it should emphasize the importance and significance of the project as it relates to the overall problem. A good summary provides all this information within one to two paragraphs.

Budget

If you have prepared a proposal for a foundation or submitted a preproposal, you already have some idea of the costs of your project and the fiscal limits of the funding source. However, most foundations and government agencies require detailed budget information. Federal agencies usually provide specific budget forms and instructions on how they are to be completed. Since the task of budget preparation is much simplified once the main body of your application is finished, it should be one of the last steps in proposal preparation.

Although it is difficult to project every essential item that may be required for your research—for the last year of a three-year grant you are essentially being asked to do this four years in advance—a standard, detailed budget page used by NIH is very helpful in this regard. While preparing the budget you should also keep in mind what aspects of this part of the proposal the reviewers will critique: the budget needs to be realistic in respect to the specific aims and the procedures by which they are to be achieved. All items should be justified on the basis of your experimental approach, the methodology employed, and the means of data analysis. Now let us turn to some specifics as we work our way through a hypothetical case.

Direct Costs. The first type of costs to be considered are those that are directly incurred by the grant's activities. These usually fall within distinct categories such as personnel, supplies, equipment, travel, and so forth.

Personnel. In table 8-4, the principal investigator has listed the names and positions of all professional and nonprofessional personnel associated with the project. He has then estimated the time or effort each is to devote to the project. Next, the dollar amount for each individual's salary, based upon the percentage of effort or time devoted to the project, has been calculated. Salaries should be commensurate with the salary scale at your institution for individuals of equivalent training and experience. Although the salaries listed will of individuals for whom support is requested, and the purpose of the travel should be listed. Per diem rates of the institution should be included as a travel expense. For foreign travel, the NIH requires a more extensive description and justification. You must relate how this type of travel is necessary to accomplish project goals.

Other expenses. This category includes miscellaneous items that are required by the project but are not attributable to other categories. Typically included are such items as reprographics, illustrations, page and reprint costs of publication, computer time, postage, and books.

Most granting sources allow for a number of other costs such as those associated with lease of office or laboratory space, minor alterations and renovations, third-party or contractual costs, and consulting fees. These items are also considered direct costs and require detailed justification. Usually strict guidelines governing claims for these expenses are outlined.

Indirect Costs. The Public Health Service defines indirect costs as those not readily identifiable with a particular project or activity but nevertheless burdened to an institution and necessary to the general operation of that institution and the conduct of its activities (Abeles 1982). Types of expenses usually considered indirect costs include those incurred in operation and maintenance of buildings, grounds and equipment; depreciation; administrative salaries and expenses; and library costs.

Although not all foundations compensate indirect costs, many do. Those that provide indirect costs usually follow a set rate established by policy of the foundation. Be certain to determine what this rate is and to what part of your budget it applies. Generally your institution's business office will make concessions if the funding sources indirect cost rate is lower than is usually requested.

If the granting source is an agency of the federal government, your institution has probably already established an indirect cost formula agreed upon by both parties through negotiation. The most common formulas are usually based upon a percentage of the total costs of salaries and wages (excluding fringe benefits) or upon a percentage of the total direct costs. The latter is the most favorable basis of indirect cost formulas as oftentimes an investigator may wish to transfer a portion of the salary category of the project's budget to another category. When this is allowed, the grantee institution's indirect costs decrease if the compensation formula has been based upon a percentage of salaries and wages.

If your sponsoring institution is a large one, it is quite likely that decentralization of its functions and activities has occurred. While the institution is still administratively responsible, the activities of the project may take place away from the central complex. Thus both on-site and off-site indirect cost formulas must be negotiated.

Tips on Writing a Fundable Proposal

Much advice to prospective applicants on how to write a successful research proposal has been proffered (see references list at the end of the chapter). Indeed, such advice is well founded and valuable. However, it must be recognized that any research proposal reflects the personality of the individual investigator and how he thinks. Therefore it is not possible to offer advice that will fit each individual situation. By now the investigator recognizes that the research plan and its organization reflects his evaluation of the relevance of the content to be included, his critical insight and his understanding of his research field, and that it must clearly convey his plans for the proposed research. In view of the fact that his judgment, originality logic, and knowledge will be evaluated on the basis of his proposal, recommendations such as "conciseness consistent with clarity," "provocative and original," while possibly sound, are of limited utility. It is human nature to delude ourselves into believing that our own proposals surely possess the qualities underscored by those providing tips to proposal writers. Therefore we feel that an examination of why proposals fail may be more revealing and potentially useful than the constructive, but timeworn cliches usually promoted in tips for successful proposal writing.

In a study conducted by the NIH, some fifty reasons were identified as the basis for rejection of research proposals ("Why Grant Proposals Are Unsuccessful" 1981). Most proposals were disapproved for more than one reason. In fact, an average of four reasons were cited for each rejected protocol. The reasons for rejection were listed under four classes: (I) nature of the research, (II) approach, (III) investigator, and (IV) miscellaneous.

Nearly a dozen reasons for proposal disapproval were cited in class I. Those that appeared with greatest frequency (see table 8-6) seem to reflect the investigator's lack of originality or failure to think through the proposed problem carefully. We can only re-emphasize that research grants are made in support of an investigation aimed at the discovery and interpretation of facts, revision of accepted theories in light of new facts, or the application of new or revised theories. Fishing expeditions, descriptive research, data gathering, and the like, simply will not do. Nor do reviewers favor proposals that sound as though

Table 8-6. Major Reasons For Grant Application Disapproval

Category	Reasons[a]
I. Nature of research	Problem was of insufficient importance, biologically irrelevant
	Experimental purpose or hypothesis was vague
	Proposal was repetitious of previous work
	Problem was scientifically premature
II. Approach	Overall experimental design was unsound or some of the techniques employed were unrealistic
	Proposal was not explicit enough, lacked detail, or was too vague
	Problems inherent to the protocol were not recognized or were dealt with inadequately
	Methods or experimental procedures were unsuited to the stated objectives
	Application was poorly prepared or thought through
III. Investigator	Lacked adequate experience for the research being proposed
	Knowledge or judgment of literature was poor
IV. Miscellaneous	Investigator needs more liaison with colleagues in the proposed field
	Overall budget was too high

[a] In order of frequency.

they could go on indefinitely—they want results! Taking a somewhat idealistic stance, Abeles (1982) states that the proposal "should represent a series of workable experiments whose results, no matter how the experiment turns out, can be interpreted in only one way." Whereas experimentalists will readily acknowledge that this is very seldom the case, proposals prepared with this thought in mind should avoid the major criticism listed under class I. The proposal, in other words, must represent "good science."

In class II, five of some two dozen major reasons for disapproval were the most common and constituted the most frequently cited grounds for rejection in any of the classes (table 8-6). You can see that, with the exception of the first and the fourth, which reflect scientific inadequacies, the remaining reasons are based upon deficiencies in written communications skills. We cannot overemphasize the fact that the proposal reflects how the investigator thinks. If it is muddled and incoherent, the reviewer may assume that this is the fashion in which the research will be conducted since the proposal is usually the only grounds upon which a judgment can be made. It must clearly represent your ideas and thoughts. Principles of effective communication, discussed in chapter 5, are as applicable to proposal writing as to any other type of scientific communication.

With respect to the major reasons for protocol disapproval in class III, we can only advise that the investigator be objective and judicious in citing only literature that provides critical insight and is relevant to the proposed problem. The same holds true for the use of preliminary data. You must provide evidence of your ability to undertake such a program and convince the reviewing panel that the project offers a good chance of success (Somerville 1982). On the other hand, the preliminary data should be just that. It is somewhat less than honest to submit a proposal for which most of the work has already been completed. When this is done, the applicant runs the risk of rejection on basis of reason 3, class I, particularly if the reviewers are acquainted with the applicant's previous work.

The remaining major reasons for disapproval cited in the NIH study are listed in the miscellaneous category in table 8-6. We have already stressed the point that the budget must be realistic and compatible with the work proposed. As long as this rule is followed and the reviewers are able to substantiate your budget figures, little difficulty should be encountered here. In regard to consultants, it is our feeling that the frequency with which the review panels have disapproved applications for lack of liaison with colleagues in a respective field is diminishing. Indeed, when reviewers have invoked this basis of rejection it often reflects a serious deficiency in the investigator's ability to convince the panel of his experience and training in a respective area, either through the injudicious use of preliminary data, poor judgement in selection of background material, or an inadequately conceived or expressed idea. On the other hand, when excessive consultations are requested, the investigator may convey the impression that he lacks the knowledge and skills to conduct activities in areas critical to the outcome of the project. Thus the prudent use of consultants on an individual research project is advised. Conversely, a multidisciplinary program project would require a wide range of collaboration.

In summary, for a research proposal to be successful, it must manifest two of the three essential ingredients—good science and good written communication skills. We will now consider the third ingredient.

THE REVIEW AND AWARD PROCESS

We have already stressed the importance of good communicative skills in preparing an efficacious research proposal. To be successful in this endeavor, however, it is important that you have a reasonable knowledge of who will read and critique your proposal. Although the review process will vary widely from those found in small foundations in which one "expert" may review your proposal and in which certainly no harm could result from being even a distant relative of the board chairman, the trend in scientific review seems to have drifted to the model established by the NIH, that is, the so-called peer review.

By no stretch of the imagination does peer review enjoy the unqualified support of the scientific community. A recent study has demonstrated that the fate of research proposals in a particular area of the peer review system was greatly dependent upon the makeup of the peer review group and that the odds for approval were no greater than could be attributed to chance (Cole 1981). Certainly those projects that fall outside the present paradigm of scientific knowledge or practice, as well as those that are innovative and imaginative enough to provide a "quantum jump" in our current knowledge, would run great risk of disapproval under the current peer review system. Nevertheless, the peer review system, as practiced at NIH, seems to provide as nearly a nonprejudicial approach to selection of meritorious and relevant research projects as any present alternatives.

The NIH system is based upon two sequential levels of review—the scientific review group known as the "Study Section" and the National Advisory Council (Eaves 1972). A typical study section consists of ten to twenty members who act as consultants and serve terms of up to four years. They are selected on the basis of their competence as independent investigators and their achievements in their respective research or clinical areas. Competence is assessed on the basis of the quality of research conducted, publications, achievements, honors, and other significant scientific activities. In addition, such factors as geographic distribution and representation by young, minority, female, and handicapped scientists are considered in the selection process. More complete criteria for selection of Public Advisory Group members can be found in *The Research Grant Program of the Public Health Service: Issues and Considerations Relative to the Ninth Report of the House Committee on Government Operations.* (U.S. Department of Health, Education, and Welfare 1968).

In addition, an executive secretary, who is a health scientist administrator on the staff of the Division of Research Grants, serves the study section by coordinating its activities and reporting the review of each application assigned to the section. The executive secretary serves as an intermediary between the reviewers and the investigators and has the further responsibility of ensuring that each application is presented to the study section in its most favorable light.

The council review constitutes the second level of the dual review process. The National Advisory Councils are composed of a dozen individuals with an equal mix of scientists and nonscientists. The scientists are considered authorities in their respective areas, whereas the lay members are noted for their interest in national health problems. Upon recommendations from the respective institute, the Secretary of the Department of Health and Human Services appoints the council members for a term of four years.

Once an application arrives at the NIH's Division of Research Grants (DRG), the review process proceeds along the following sequence. The assistant chief of referral of the Research Grant Review Branch, or his designee, scans the proposal and makes the initial decision as to the general category or discipline into which the application falls. At this point one can see how important it is that the title and abstract accurately reflect the substance and purpose of the proposal. Based upon this initial screening, the application is then sent to a referral officer, who scans and then assigns it to the appropriate study section. The referral officers are usually executive secretaries of the various study sections and have responsibility for several disciplines. Once the application is assigned to the study section, the executive secretary reads the application to determine whether additional information is required from the applicant. He also selects, as primary reviewers, at least two of the study section members best qualified to evaluate the proposal. The proposal is mailed to these reviewers, who prepare a detailed critique that is then returned to the executive secretary prior to the next scheduled, formal meeting of the study section. This formal meeting will last two to three days, during which each proposal will be discussed. The basic points upon which the scientific merit of an application will be judged have been summarized as follows: importance of the proposed research problem; originality of the approach; training, experience, and research competence, (or promise thereof) of the investigator; adequacy of the experimental design; suitability of the research facilities; and appropriateness of the budget to the work proposed (Eaves 1972). After the primary reviewers have presented their critiques, the other members of the study section, all of whom have read the application, provide their critical commentary. A vote is then taken for one of three options—approval, disapproval, or deferral. The recommendation of the study section is by majority vote. If the recommendation is for approval, each study section member assigns a numerical score to the application based upon his opinion of its scientific merit. The mean of these scores serves as a guide to the National Advisory Council.

The recommendations of the study section are summarized by the executive secretary immediately after the panel has met. The summary statement, or "pink sheet," represents a synthesis of the written critique of the primary reviewers and the discussion of the application that occurred at the time the protocol was considered. This information, including priority score and budgetary recommendations, is then transmitted to the responsible National Advisory Council, which usually meets six to eight weeks after the study section. The council then reviews the application against a broad range of considerations such as the degree of relevance of the proposed research to the mission of the institute, the need for initiation of research in new areas, a determination of the total funding pattern of research in universities and other institutions of the scientific community, as well as the overall needs of the NIH. Although the priority score and recommendations of the study section remain the primary deciding factor for an application, the council can, based upon the preceding considerations, recommend that a proposal be placed in a category either to be funded or not. The latter decision is most frequently based upon the programmatic relevance of the application to the goals of the institute.

Once a decision by the council has been reached, the administrative staff of the institute must then match the approved applications, by priority score, to the funds available. Here the priority score becomes a critical factor. The successful grantee is then notified by a "Notice of Grant Award." This document delineates the terms of the award, including starting date, grant period, amounts budgeted, and any special administrative recommendations or conditions. The path of an application through the NIH review process is represented in figure 8-2.

On occasion, action on an application by the study section may be deferred. The usual reason is that further information is required before a decision can be reached. We have already indicated that one responsibility of the executive secretary is to act as intermediary between the applicant and the study section. When this responsibility has been fulfilled, then any additional information required by the study section is of a nature that cannot be obtained by usual means of communication, and an on-site visit is warranted. Often, the site visit is made when large budgets are involved or when unusually long periods of support are requested. Merritt and Eaves (1975) have found that the type of information most frequently sought via the site visit includes the extent of the principal investigator's experience and knowledge about the respective research area or commitment to the proposed research project; the nature and characteristics of collabo-

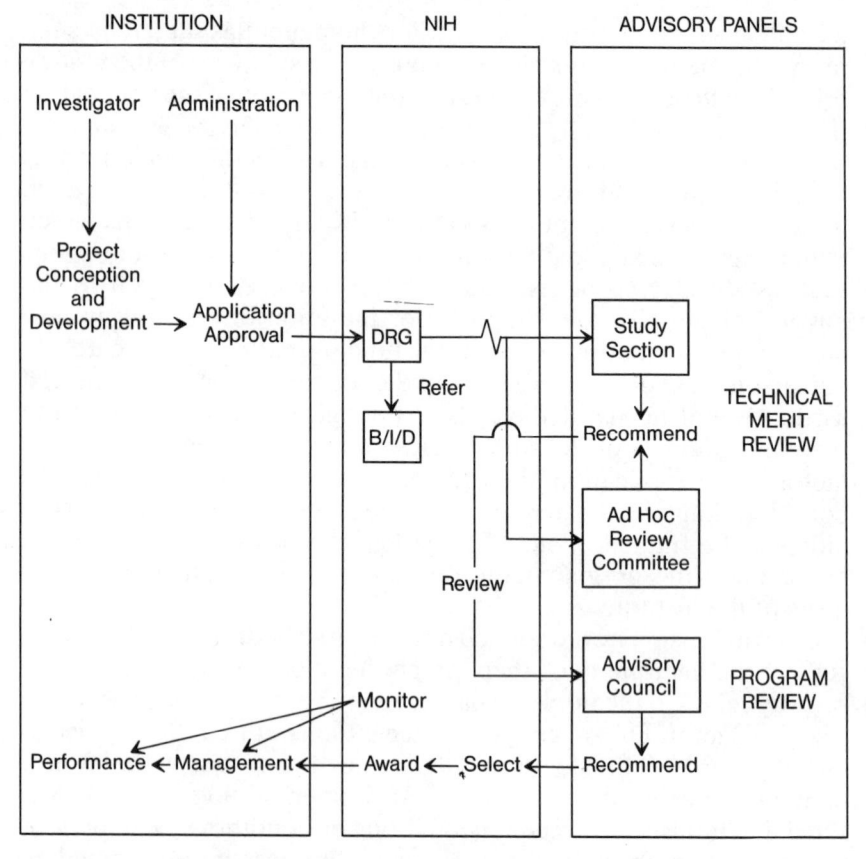

Figure 8-2. The path of a grant application from conception through the NIH review process. DRG is the Division of Research Grants, B/I/D is Bureau/Institute/Division.

rative arrangements; progress made to date on the project; nature of the physical facilities including laboratory space and equipment; and institutional interest, support, and commitment, as well as first-hand observation of special techniques, equipment, or procedures essential to the proposed project.

It is usually the executive secretary or a representative of the agency who arranges, with the principal investigator, a suitable date for the site visit. At this time the executive secretary discloses who will compose the site-visit team; and an informal agenda, containing the salient points raised by the study section, is developed. Usually the visitors will want to interview the coinvestigator and other professional personnel associated with the project, and it is advisable to have the institution's administratively responsible official available for interview. The site visit for a project grant application rarely lasts longer than one day and is considerably less formal than that for a program project. Although it would be improper to favor the site-visit team with amenities such as meals or lodging, it is appropriate to arrange transportation, provide information concerning convenient, comfortable lodging, and assist in any way that will make their stay pleasant.

When an applicant has competed unsuccessfully, whether the proposal is disapproved or approved but not funded, profit can still be gained from the experience. First, the "pink sheet" should prove helpful in emphasizing the areas the reviewers felt were weak. In addition, the executive secretary will often be able to provide additional information from notes taken at the time the application was being considered by the study section. By using this information, the applicant is in a much stronger position to revise the application or reorganize for a future submission.

The successful application must reflect at least three critical ingredients. Good science and written communicative skills have already been stressed. Because of the impact on success or failure that the constituency of the study section imparts, the third essential ingredient must be *good luck!* Regardless of the shortcomings inherent in the peer review system, however, we must remember that it does entrust the quality of science in this country to the guidance and surveillance of the scientific community itself. Most reviewers accept this commission seriously, and when the system fails, we have only ourselves to blame. With this sobering thought in mind, Eaves (1972) advises, "An author cannot, then, in good conscience do less than his best in attempting to prepare a complete and lucid application that will serve his own interests as faithfully as his peers serve him."

GRANT MANAGEMENT

Once an award has been made, the principal investigator should not be content with discharging only those responsibilities associated with the technical direction of the project. Whereas the granting agency usually makes the award to the institution on your behalf and despite the fact that the institutional official who signed the application has assumed de facto responsibility for the administrative management of the grant, you as principal investigator should at least be aware of the responsibilities and conditions to which the institution has agreed. Certain requirements of accountability are imposed on federal discretionary grant activities where public monies are concerned. When the investigator is located at a small university or institution, the business office is not likely to employ a restricted funds manager and may have little experience with grants management. Even in large institutions it is well to monitor the management of your grant to assure compliance with the accounting requirements and specified regulations of the sponsoring agency.

It has been suggested (Kenny 1980) that the necessary administrative responsibilities of the principal investigator generally fall into two categories. The first is an internal function in which the investigator must assure that the management of the grant be integrated into the accounting and business system of the institution. At institutions where the volume of grants is small, the investigator will need to become familiar with the fund accounting system. Without experienced grants management personnel, the development of a line item budget ledger will help avoid unnecessary audit problems later. Even so, unless fund expenditures strictly adhere to the line item budget, which is seldom the case, changes in expenditures commonly occur. This is seldom a major problem in institutions where large numbers of grants are routinely managed. Even in this instance, however, it is advantageous to the investigator to monitor closely the activities of the business office or grant management section as they relate to administration of his grant.

A simplified managerial system employing *posting documents* and *financial source summaries* that are useful for this function has been described by Black and Starr (1977) (fig. 8-3). The posting document identifies the financial source or grant account number. This document is used to record supplies receivable, to provide a record of partial shipments or items on back order and a running estimated category balance, and to identify each purchased item as to requisition and purchase order number. Similar posting documents can be maintained for personnel, equipment, or any other cost-category budgeted line item. At the end of an accounting period, usually each month, the investigator can use the financial source summary to review cost-category deviations, remaining net funds available for each cost category, and balances in each major fund source, as well as the laboratory's total funding. To accomplish this the investigator must first compare the estimated cost of each item of final balance, by purchase order and requisition number, to that shown on the institution's computer records. Deviations of the actual costs from the

GRANTS AND GRANT MANAGEMENT 163

Posting Document

SOURCE OR GRANT NUMBER __4321-04__   ACTIVITY __SUPPLIES__   PREVIOUS BALANCE __$6,000__

	Date	Req.	PO	Item	Quan	Vendor	Est. Cost	Est. Balance	Date Rec'd	Adj. Cost	Adj. Balance	Remarks
(A)	1-10-8X	39406		Disposable Pipettes	6(cs)	Ajax Sci. Co.	$780	5,220				
(B)	1-10-8X	39406	96401	Disposable Pipettes	6(cs)	Ajax Sci. Co.	$780	5,220	1-24-8X	$800	5,200	Catalog price inc.

NET ADJ. BALANCE __$5,200__

Financial Source Summary

SOURCE OR GRANT NUMBER __4321-04__   END REPORTING DATE __1-30-8X__
                                               TERMINATION DATE __12-30-8X__

Category	Total Budgeted	Actual Experience this month	Actual Experience to date	Total Encumbered	Net Available	Standard Budgeted	Variances	Remarks
Personnel	$10,000	$833	$833	$10,000	-0-	-0-	-0-	Salaries encumbered for entire activity period
Supplies	$6,000	$800	$800	-0-	$5,200	$500	<$300>	
Equipment	$2000	-0-	$1,800	–	$200	–	–	–
Travel	$500	-0-	$200	–	$300	–	–	–

Figure 8-3. Internal laboratory accounting documents. The Posting Document is used to track laboratory expenditures. The Financial Source Summary provides a review of cost-category deviations, remaining net funds available for each cost category, and balances for any specified accounting period.

expected costs usually result from differences in shipping charges, catalog price changes, or accounting errors in the business office. In case of error, the accounting system provides a check against the business office so that only traceable costs are charged to the laboratory. Unjustified deviations must then be adjusted in the business office. When justified, deviations should be used to compute an adjusted cost balance on the posting document. Next, the adjusted balances from each category are brought forward to the financial source summary. Computations of total amount budgeted minus total expenditures and encumbrances yield net available—or net deficit—for the rest of the funding period. An internal accounting system in the laboratory provides a check on the institution's accounting performance and gives financial control over both individual projects and activities as well as the laboratory's total resources. It also allows a measure of performance evaluation and creates a basis for future planning. With the advent of the small personal computer (PC), a system of accounting suitable for this function is available to almost everyone. Indeed, menu-driven commercial programs that address nearly every aspect of grant management are now available for use with the PC.

While internal accounting responsibilities can be handled in the fashion just described, the external accounting responsibilities require strict compliance with the guidelines and regulations of the sponsoring agency. Some specific aspects of grants management, as required by NIH, follow (U.S. Department of Health and Human Services 1986).

Changes in Expenditures, Allowable Costs, and Prior Approval

Once a project is underway, the recipient may find it difficult, if not impossible, to restrict expenditures within the budget categories of the approved project budget. The NIH allows grantee institutions a certain degree of latitude for rebudgeting within and between budget categories in order to achieve programmatic goals. However, the grantee must aquire prior approval from NIH for the following conditions:

- The award of additional funds (not including categorical changes that increase the base upon which indirect costs are calculated)
- Expenditures that have been expressly disapproved or restricted as a condition of the award (including travel funds)
- Transfer of indirect cost monies to direct cost categories

- Transfer of monies between construction and nonconstruction categories
- Transfer of trainee costs to other budget categories
- Any expenditures for general support services of $25,000 or 10 percent of the total direct-cost budget, whichever is greater

In order to obtain prior approval for cost transfer or rebudgeting, the grantee must submit the request in writing to the grants management officer designated on the "Notice of Grant Award." The request must bear the signature of both the principal investigator and the authorized institutional official. These requests are reviewed by the responsible grants management officer, who then informs the grantee of the final disposition of the request. It is important that the grantee assure that the letter of disposition is signed by the *authorized* NIH official; otherwise the response is not considered binding. If you are uncertain about allowable costs or types and levels of expenditures, it is wise to consult your grants management officer. Failure to obtain prior approval when required may result in disallowance of the incurred costs.

The Public Health Service (PHS) has the option of conditionally waiving the prior approval requirement for certain types of costs and grantees. Grantee institutions under this option include private nonprofit organizations, state and local government agencies, hospitals, colleges, universities, and research institutes and foundations. These organizations must institute their own prior approval system following specific PHS standards. Although institutional prior approval systems are subject to PHS review and audit, the local system introduces a greater degree of flexibility with a faster response time and thus provides a more efficacious path for pursuing project objectives.

Reports of Expenditures

The NIH routinely monitors their grants to assure that programmatic goals are reached and project funds are properly spent. The latter is ascertained through reports of estimated and actual expenditures. A report of estimated expenditures is required with submission of the grantee's annual progress report or continuation application. The grants management officer carefully monitors the grant's financial status. When the reports of estimated expenditures indicate that an unobligated balance will exist at the end of a budget period, one of the following courses of action may be taken:

- The estimated unobligated balance may be applied to the following year of a continuation award. In effect, the unobligated balance is subtracted from the federal share of the approved, budgeted amount for the continuation year.
- The unobligated balance may be used to increase the level of funding in the continuation year. This, of course, requires justification by the grantee and NIH approval.
- The unobligated balance may be withdrawn from the current award authority.
- Small amounts (currently $250 or less) in the estimated unobligated balance can be ignored.

A report of actual expenditures, the *financial status report*, must be submitted ninety days after the close of the budget period. Where actual expenditures are less than the NIH share of the approved budget, leaving an unobligated balance, the grants management officer has the following options:

- Issuing a revised award notice for the current budget period and withdrawing the excess funds
- Upon written request, allowing the grantee to expend the excess funds for approved project purposes
- Disallowing the expenditure of excess funds in the current budget period and applying the balance to the budgeted amount for the continuation year

Unobligated funds at the end of a *project period* revert to the federal agency. If the actual expenditures exceed the budget amount by more than $250, the grants management officer may either make an administrative supplemental award if funds are available for obligation or negotiate with the grantee a satisfactory revised budget to make the NIH share of the budget equal to the funds available. In the latter case the grantee must then be accountable for making up the short-fall from other nonfederal funds.

Progress Reports

The programmatic performance of the project is monitored by NIH through annual progress reports that must be submitted with a competing, or noncompeting continuation application. These reports are due sixty days prior to the start date of a continuing grant period or included in new and renewal competing applications. In addition, a terminal progress report is required upon expiration of a grant.

Normally these reports require a statement of the project's general scientific goals; a description of the studies conducted during the current grant period, the results, and their significance; and specific objectives for the coming grant period. In the past, NIH has required a list of publications resulting from the project.

Invention Reports

In addition to programmatic performance and expenditure records, all inventions made under the aegis of NIH support must be promptly and fully reported. This requirement includes training grants, traineeships, and scholarships. Furthermore, each application for competing or noncompeting continuation support requires either a declaration of all inventions made during the preceding budget period or certification that no inventions were made. Upon expiration or termination of a grant, a *final invention statement and certification* must be submitted within ninety days. In this statement, all inventions conceived or actually reduced to practice during the period of project support must be reported.

Transfers, Termination, and Special Circumstances

Transfer of Grants. When a principal investigator transfers employment from one organization to another, the project grant that he directs may also be transferred without competitive review when the following conditions are met:

- The original grantee institution agrees to transfer the project by submitting a statement relinquishing its interest.
- The new institution submits a new application form declaring administrative support of the project.
- The principal investigator plans no significant changes in direction of the research or level of funding from that originally approved.
- The facilities and resources at the new institution are adequate for successful performance of the project.

When these conditions are met, the project may be transferred to the new institution for a period up to the remainder of the previously approved project period in an amount not exceeding that recommended for the remaining period. If these conditions are not met, the awarding office may then require competitive review of the project.

Audit and Audit Resolution. We have mentioned the means by which the NIH routinely monitors the programmatic progress, records of expenditures, and invention status of a funded project. However, the financial operations, including reports, resources expended and managed under the terms of the grant, and efficacious achievement of project objectives, may all be the subject of a formal systematic review, or *audit*. Any or all of these elements may be audited at the discretion of the federal government either during or after the period of NIH support. Most frequently these audits relate to the financial operations of the grantee institution and specifically to the allowability of costs burdened to grant-supported activities. Grantees are allowed thirty days to respond to the audit resolutions officer concerning audit findings. Failure to do so may automatically result in disallowance of costs. A systematic process for resolution of audit findings, between institutional officials and the audit resolution officer, has been established. At the completion of this process, the grantee is notified of the final decision. If the grantee feels the decision is unjustified, an appeals procedure is available whereby the grantee submits a request for review. A review committee, composed of officials not involved in the original adverse decision, will rule on the case. If the grantee remains unsatisfied with this decision, a request may be submitted to the executive secretary of the departmental grants appeals board for further review.

SUMMARY

A grant is a mechanism by which financial assistance is provided to an organizational entity or individual to support approved activities. Although there are numerous institutions that finance grants, the federal government is by far the most wealthy—at least in terms of scope and dollars spent in support of research activities. The federal research project grant has served as the basis for our discussion of the principles of grantsmanship and grants management although these same principles for the most part are applicable to privately supported grants as well.

Throughout the process of applying for grant support, the pre-application stage is of vital importance in deciding the outcome of your quest. You must satisfactorily answer the following questions: Are you certain that the idea you are proposing is novel or innovative and responsive to some important and timely issue or problem? Do you have the background, experience, and expertise to carry the project through to fruition? Will your parent institution provide the support and environment necessary to bring the project to a successful conclusion? If these questions can be satisfactorily answered, your next step will be to identify potential sources of funding. These funding sources must share an interest in your ideas or the questions you have posed. The project goals must be compatible with those of a potential grantor. Further refinement can be achieved by screening the list of potential funding sources for application submission deadlines and review criteria. Adequate time to prepare a polished proposal is essential, and it is important that you have some knowledge of the evaluation criteria used in judging your proposal.

Once the source with greatest potential for funding has been identified, the most effective written communicative skills must be employed to convince that source that the applicant is the individual most capable of successfully accomplishing the project's goals. To this point, both *good science* and *good communicative skills* are requisite ingredients for successful grantsmanship.

Finally, the applicant must know something about the type of individual or the makeup of the panel that will review the proposal. Whereas knowledge of the review criteria is an important factor in the selection of a potential grantor, the knowledge of the type of individual that will review your proposal is essential if you are to communicate your ideas effectively. However, the circumstances of the actual review represent elements over which the applicant has no control. Thus the third essential ingredient of successful grantsmanship is *good luck!*

After an award is made, the investigator not only has the responsibility of directing the technical and scientific aspects of the project but must actively participate in the administration and management of the financial resources as well. Where federal monies are involved, certain requirements of accountability are imposed upon the management of such funds. In addition, there may be specific conditions imposed upon the parent institution in management of the grant. The investigator should be aware of any restrictions or conditions and assure that both his and the parent institution's responsibilities are met in a timely, accurate, and efficient manner.

REFERENCES

Abeles, F. B. 1982. "How to Prepare an Effective Scientific Research Proposal." *The Grantsmanship Center News* January/February, 10(1):52-57.

Black, H. S., and D. A. Starr. 1977. "Nonprofit Lab Managers: Ease Your Accounting Woes." *Laboratory Management* 15:48-50.

Cole, S., J. R. Cole, and G. A. Simon. 1981. "Chance and Consensus in Peer Review." *Science* 214:881-86.
DeBakey L., and S. DeBakey. 1978. "The Art of Persuasion: Logic and Language in Proposal Writing." *Grants Magazine* 1:43-60.
Eaves, G. N. 1972. "Who Reads Your Project-Grant Application to the National Institutes of Health?" *Federation Proceedings* 31:2-9.
_____. 1978*a*. "A Successful Grant Application to the National Institutes of Health: Case History." *Grants Magazine* 1:263-83.
_____. 1978*b*. "The Research Grant Application: An Exercise in Scientific Writing." *Grants Magazine* 1:61-63.
Gee, H. H. 1973. "Preparation of the Project-Grant Application: Assistance from the Administrator in Charge of a Study Section." *Federation Proceedings* 32:1544-45.
Jagger, J. 1980. "How to *Write* a Research Proposal." *Grants Magazine* 3:216-22.
Kenny, J. T. 1980. "The Accounting Responsibilities of the Project Administrator." *Grants Magazine* 3:227-31.
Kiritz, N. J. 1979. "Program Planning and Proposal Writing." *The Grantsmanship Center News* May/June, 5(3):33-79.
_____. 1979. "Proposal Checklist and Evaluation Form." *The Grantsmanship Center News*, July/August, 5(4):42-45.
Merritt, D. H. 1963. "Grantsmanship: An Exercise in Lucid Presentation." *Clinical Research* 11:357-77.
Merritt, D. H. and G. N. Eaves. 1975. "Site Visits for the Review of Grant Applications to the National Institutes of Health: Views of an Applicant and a Scientist Administrator." *Federation Proceedings* 34:131-36.
Murry, J. P., and B. R. Biles. 1980. *Tips for Proposal Writers.* Manhattan, Kans.: Graduate Services & Publications.
Pike, J. M., and S. C. Bernard. 1978. "The Research Grant Budget: Preparation and Justification in Relation to the Proposed Research." *Grants Magazine* 1:283-86.
Schimke, R. T. 1973. "Preparation of the Project-Grant Application: Assistance from the Grantee Institution's Experienced Investigators." *Federation Proceedings* 32:1548-50.
Somerville, B. 1982. "Where Proposals Fail." *The Grantsmanship Center News* January/February, 10(1):24-25.
Stallones, R. A. 1975. "Research Grants: Advice to Applicants." *Yale Journal of Biology and Medicine,* 48:451-58.
Teague, G. V. 1981. "Request for a Proposal: Solicitation for a Federal Contract." *Grants Magazine* 4:16-28.
U.S. Department of Health, Education, and Welfare. 1968. *The Research Grant Program of the Public Health Service: Issues and Considerations Relative to the Ninth Report of the House Committee on Government Operations.*
U.S. Department of Health and Human Services. 1986. *Grants Administration Manual.* Bethesda, Md.: U.S. Department of Health and Human Services.
U.S. Public Health Service. *PHS Grants Policy Statement.* DHHS Pub. no. (OASH) 82-50,000. Washington, D.C.: U.S. Government Printing Office.

White, V. P. 1975. *Grants: How to Find Out About Them and What to Do Next.* New York: Plenum.
"Why Grant Proposals Are Unsuccessful." 1981. *Health Grants and Contracts Weekly,* March.

GENERAL BIBLIOGRAPHY

Catalog of Federal Domestic Assistance. 19th ed. 1985. Washington, D.C.: U.S. Government Printing Office.
Commerce Business Daily. Washington, D.C.: U.S. Government Printing Office.
Council on Foundations. *Foundation News: The Journal of Philanthropy.* (Published bimonthly.) New York: Council on Foundations.
Directory of Federal Health & Medicine Grants and Contracts Programs. 1979. Bethesda, Md.: Science and Health Publications.
Foundation Center. 1975-76. *The Foundation Center Source Book.* New York: Foundation Center.
_____. 1979. *The Foundation Directory.* Edited by M. O. Lewis. 7th ed. New York: Columbia University Press.
Foundation Directory. (Published biannually.) New York: Foundation Center.
The Foundation Grants Index. (Published annually.) New York: Foundation Center.
Government Information Services. *Federal Funding Guide.* (Published annually.) Washington, D.C.: Government Information Services.
Grants Magazine. (Published quarterly.) New York: Plenum Press.
The Grantsmanship Center. *Grantsmanship Center News.* (Was a bimonthly publication; no longer published.) Los Angeles.: The Grantsmanship Center.
National Institutes of Health. *NIH Guide for Grants and Contracts.* (Published weekly.) Bethesda, Md.: Office of Grants and Contracts, NIH.
National Institutes of Health. 1985. *Guide for the Care and Use of Laboratory Animals.* DHHS (NIH) Publication no. 85-23. Bethesda, Md.
National Science Foundation. *NSF Grant Policy Manual.* 1983, revised ed. Washington, D.C.: U.S. Government Printing Office.
U.S. Department of Health and Human Services. *Guide to NIH Programs and Awards.* Bethesda, Md.: U.S. Department of Health and Human Services.
U.S. Department of Health and Human Services. *Grants Administration Manual.* 1986. Bethesda, Md.: U.S. Department of Health and Human Services.

as one requiring acquisition of a large animal colony, an initiation fee is sometimes included as part of the payment specifications in the contract.
- The awarding agency, private or governmental, generally exercises tighter financial control over contractual studies than those covered by grants.

There are other, more subtle, differences between grants and contracts that are issued by governmental agencies (Kooi 1978). These differences will be discussed in more detail later in this chapter.

CONTRACTING TO DO RESEARCH

Types of Contracts

Several types of contracts can be solicited by both federal agencies and private research organizations.

Fixed Fee. When a fixed fee contract is signed, the contractor agrees to perform a specified task for a fixed amount of money. Typically, toxicology studies done for the private sector are negotiated on a fixed fee basis. It is extremely important for the contractor to have tight control over his costs to assure that he makes a profit on a fixed fee contract.

Cost Plus. A safer type of contract for the contractor is the cost-plus fixed fee contract. In this case, the contracting party will pay all costs incurred by the contractor that are specifically included in the contract plus a fixed fee. However, the awarding agency will probably exercise very tight accounting procedures on the contractor and will also probably insist on periodic audits of all expenditures to minimize the possibility of cost overruns.

A variation of the cost-plus contract is a contract based upon the number of labor hours required to complete the study. In this case, the contractor specifies a fixed cost per hour that includes both his direct and indirect costs and his internal profit factor.

Time Plus Materials. Another variation is the time-plus-materials contract. This type of contract may be used where special materials are required that are not part of the standard laboratory inventory. An example would be a metabolism study requiring the purchase of a customized radiolabelled compound to be used in the conduct of the study.

Bid Solicitation Procedures

Bids are made through prespecified procedures. Some contracts are awarded on the basis of competitive bidding procedures, others on sole-source procurement. The latter is used if the awarding agency or private organization believes that only one prospective bidder has the capability to conduct the required study. Sole-source procurement is also used for continuation of a particular study in which the contractor's experience is an obvious advantage. Contracts also may be awarded based on a restricted bidding procedure whereby only those organizations selected by the contractee are solicited to submit competitive bids for the contract. The contractee may "prequalify" potential contractors by issuing a request for information (RFI) setting forth the nature of the services required, proposed budget, time frame, and performance specifications. Usually the contractee will require certain information regarding the qualifications of potential bidders who respond to the RFI. Federal contracts in the biomedical sciences are often restricted geographically—for example, to the area within a 150-mile radius of Bethesda, Maryland, the location of the NIH. This restriction is common when the contract is part of work being conducted in house and if a high degree of monitoring is required.

Competitive bids are usually solicited when the contracting party believes that a better price or higher quality work can be obtained by making the contract available to many sources. The awarding agency may open the competitive bidding procedure in any of three different ways: through a request for proposal (RFP), a request for quotation (RFQ), or an invitation for bid (IFB). Although the differences between these three solicitation vehicles may seem obscure, (the terms RFQs and IFBs are often used interchangeably), they can be clarified on the basis of the detail and specifications required in each case.

RFPs. An RFP is actually a request for a solution to a problem and thus requires a high degree of technical creativity on the part of responding organizations. In addition to costs, the agency is interested in the approach that the contractor will use to conduct the study. The RFP will include background information and current activity so that potential bidders may develop their own protocols to conduct the study.

RFQs. Requests for quotations are another type of solicitation that requires technical creativity. In this case, the soliciting agency will provide a general description of the required services and include an outline of how the agency feels the contractor should provide those services.

IFBs. Invitations to bid are the most specific form of solicitation for contractual work. The IFB specifies not only the work to be completed, but details how it is to be performed. These parameters, along with specifications, necessary reports, the reporting time frame, and the completion date of the project, leave little technical creativity to the discretion of potential contractors. The final decision is generally based on price, with the contract going to the lowest qualified bidder.

The thoroughness with which a potential contractor responds has a great deal of influence on his eventual success. Prior to making application for a contract, the investigator should make sure that his application or bid is prepared exactly according to the agency's requirements. The application may otherwise be rejected out of hand and not given the consideration it deserves. Success is also affected by the type of contract you are trying to obtain: it is sensible to concentrate on contracts for which your organization has the advantage of being the only acceptable source or of being included in a restricted bidding classification.

Contractual Opportunities

The investigator seeking information on contracts to be awarded may consult several different sources. A government publication, *Commerce Business Daily (CBD),* provides listings of contracts. Private publications such as *Federal Contracts* and *Grants Weekly,* published by Capitol Publications, are also useful in this regard.

Decision Criteria

Prior to requesting a copy of a RFP, your organization should critically examine this type of funding before deciding to pursue it. Federal agency contracts involve several stringent requirements that do not apply to grant funding.

The first of these is the time frame in which you must complete and submit your proposal to the agency. Typically, the proposal must be received by the agency within thirty to forty-five days after publication in the *CBD*. In some cases, an even shorter response time, such as twenty days, may be specified. Other factors that may influence your decision to seek federal agency contract funding are the more stringent auditing and reporting requirements associated with contract studies. The major questions to be answered prior to requesting a copy of the RFP are the following:

- Is our organization interested in obtaining contractual funding?
- Does our organization have the general expertise currently available to handle the type of work described in the proposal announcement?
- Do we have internal accounting and reporting procedures that will satisfy the awarding agency?
- Is there sufficient time for us to prepare a proposal and to submit it to the agency within the time constraints?
- Are there any restrictions listed in the announcement that would exclude our organization?

Since the costs of preparation of the proposal can be substantial (one estimate developed in 1980 suggested an average cost of approximately $2,000 for proposal development) (Hensley, Gulley, and Edelman 1980), and since these costs are generally not recoverable in the contract, only those announcements that satisfy the above criteria should be pursued.

Preparation of the Proposal

Upon receipt of the RFP, a more thorough reading of the request is required to determine if the investment of effort in the preparation is still warranted. You must be satisfied that your organization has a reasonable chance of being awarded the contract if you submit your proposal. In some cases, the awarding agency will schedule a bidder's conference to answer questions regarding the solicitation. This conference will not only give you an opportunity to obtain additional information regarding the proposal, but will also give you an indication of the other organizations that are interested in submitting a proposal. If the current solicitation is a continuation of a previous contract, you may not want to respond, since the holder of the original contract will have a distinct advantage over other bidders.

The format for RFPs is not standardized among the various agencies using this procedure. However, major headings are likely to include scope of the work, review procedures and evaluation criteria, and proposal preparation requirements. One experienced in making applications for grants will observe certain similarities in the preparation requirements for both grants and contracts.

The *scope of the work* section defines the tasks you propose to complete if you are awarded the contract. Obviously, your response should closely parallel the requirements listed in the RFP. Your response should show that you have thoroughly evaluated the project, anticipated potential problems, and considered alternate approaches.

The *review procedures and evaluation criteria* section should be thoroughly reviewed to ascertain that your proposal is favorably weighted to take advantage of the criteria established by the agency. The relative importance of criteria varies among solicitations and among the agencies making those solicitations. Your consideration of the relative importance of these judging criteria may be one of the most important factors in determining your success in obtaining the contract.

The *proposal preparation requirements* section specifies the organization of the information to be included in the proposal. If an outline is included in the RFP, you should follow it exactly in writing and presenting your proposal. The required number of copies of both the technical proposal and the budget information should be packaged and submitted to the agency by the required submission date. It may be necessary to deliver it by hand or use the services of a guaranteed delivery service to make sure that your proposal is reviewed.

Contract Management

The principles of grant management discussed in chapter 8 apply to the management of contract studies. However, because of differences in auditing procedures that apply to contractural studies, your accounting personnel and system must be able to adjust to the conditions set forth in the particular contracts awarded.

SPONSORING CONTRACTUAL RESEARCH

Contracts in the Private Sector

A significant amount of contract research is conducted by organizations in the private sector. In many cases a particular study that is required by a regulatory agency for product registration must be contracted due to lack of qualified personnel or facilities within the private sector organization. Examples of this type of work include toxicology; environmental or large animal metabolism studies required under the provisions of the Federal Insecticide, Fungicide, and Rodenticide Act (FIFRA); or studies needed to obtain clearance for a cosmetic product under the jurisdiction of the FDA. This type of research is not restricted to private contract laboratories. A great deal of the contract work conducted for private industry is conducted in universities by principal investigators employed by the academic institution.

Both the organization requiring contractual research and the organization proposing to provide the contractual research are faced with problems in the awarding and management of this type of research. We will attempt to identify the types of problems that can arise and to propose some solutions to avoid the pitfalls of contract research.

The Awarding Organization's Concerns

The awarding of contractual research studies depends on several factors, of which cost may be the least important. Many organizations have procedures—ranging from informal to highly structured—that govern the selection of outside laboratories to conduct research studies for them (Teague 1981). Whether the procedure is formal or informal, there are several major concerns that the awarding organization must satisfy prior to awarding the contract. These would most certainly include, at a minimum, the following:

- Can the bidding organization provide the type of research we require?
- Will the contract lab conduct the research using protocols prescribed by regulatory agencies such as the EPA under the provisions of FIFRA or the Federal Hazardous Substances Act (FHSA)?
- Will the contract lab deliver the results of the research according to our timing requirements?
- Does the contract laboratory have a reputation for delivering high-quality work?
- If there are several laboratories capable of conducting the required research, should the contract be put out for competitive bid?
- Does the contract laboratory have the appropriate accreditations to conduct this research?

Selection Procedures

If your organization does not have a formal procedure for the selection of outside contract laboratories, the above questions may pose some problems for you. There are at least five phases in formal selection procedures used in the industry. These might include:

- Identification of outside laboratories
- Preliminary evaluation
- Site visit
- Trial study
- Decision

Identification. The identification of outside laboratories that are capable of conducting the desired research may be the most informal of the steps involved in selection of contract labs. One common way of locating outside labs is through personal contact at scientific meetings of the Society of Toxicology, Weed Science Society of America, Plant Growth Regulator Working Group, the Society of Cosmetology, or similar professional groups. Another source of potential laboratories are references such as the *International Directory of Contract Laboratories* (Jackson 1985), which lists various contract laboratories and their capabilities.

Preliminary Evaluation. Prior to serious consideration of a particular laboratory, several evaluations should be made. These include a review of the protocols used by the laboratory, examination and comparison of a price list with that of similar laboratories, a review of studies previously conducted by the laboratory (there may be confidentiality agreements to be considered), a verification of accreditation, examination of the curriculum vitae of key personnel, and a review of procedures for invoicing and payment. Other factors that might influence your selection include the ownership of patent rights, the contract laboratory's ability to design a study to meet your requirements and to develop and validate new methods or modify existing methods.

Site Visit. After a potential contract laboratory has successfully passed the preliminary evaluation, it is advisable to inspect physical facilities and observe the manner in which work is conducted at the laboratory. This site visit should be conducted by the personnel who will be evaluating the results of the studies to be contracted.

Trial Study. Probably the best way to evaluate the work of a contract laboratory is to schedule a trial study that has been previously run at an approved laboratory. This trial also gives an indication of the responsiveness of the contract laboratory in meeting scheduled deadlines and provides an opportunity to run through billing and payment procedures. After the report is received, it should be thoroughly examined for both accuracy and completeness. If there are discrepancies between the trial study report and the original study report, they should be resolved prior to the sponsor's acceptance of the report.

If the contract laboratory has met all of your requirements and is acceptable either as a primary or secondary source of a particular study, the contract laboratory should be notified of its acceptance and be included on subsequent RFQs.

CONTRACTUAL RESPONSIBILITIES

Sponsor's Responsibilities

Both the sponsoring organization and the contract laboratory have responsibilities in conducting contract research. The sponsor should review only those laboratories that have a reasonable chance of being selected to conduct work for the sponsor. The sponsor should make sure that the contract lab clearly understands the sponsor's time requirement, reporting requirements, and required protocols. The sponsor should also provide the contract lab with as much information as is possible regarding the handling of materials to be used in the study to minimize risk to contract lab personnel from potentially toxic materials. In the case of a material that is being submitted for initial toxicology testing, the sponsor should provide as much information as possible regarding the chemical class and/or structure of the compound.

Contractor's Responsibilities

The contract laboratory also has responsibilities to the sponsoring organization. The most important of these is to conduct the study using good laboratory practices that will insure the scientific creditability of the research. In some cases, regulatory agencies have published *good laboratory practice* (GLP) guidelines or regulations that must be followed while conducting the research. Included under GLP requirements is the responsibility of the contract lab to maintain complete records regarding each study conducted, should a question be raised regarding the adequacy or accuracy of a study at some point in the future. The contract lab also has a responsibility to maintain complete confidentiality regarding the studies being conducted by a particular sponsor.

SUMMARY

Use of contractual laboratories by both federal agencies and private industry is becoming even more important in the light of increased regulatory pressures and the demand of the public for solutions to pressing problems of national concern.

While the procedures involved in the use of contract labs can be complex, a thorough understanding of these procedures can lead to increased productivity for the sponsor of the research and profitability or increased sources of funding for the organizations conducting the contractual work.

REFERENCES

Hensley, O., B. Gulley, and J. Edelman. 1980. "Evaluating Development Costs for a Proposal to a Federal Agency." *Journal Society of Research Administrators* 12:35-39. 12:35-39.

Jackson, E. M., compiler. 1985. "How to Choose a Contract Laboratory: Utilizing a Laboratory Clearance Procedure." *International Directory of Contract Laboratories.* New York: Marcel Dekker.

Kooi, B. Y. 1978. "Analysis Grants and Contracts... Same or Different?" *Health Grants and Contracts Weekly,* September 7, 9-10.

Teague, G. V. 1981. "Request for a Proposal Solicitation for a Federal Contract." *Grants Magazine* 4:16-28.

Chapter 10
Compliance with Governmental Regulations

In the past, laboratories were not stringently regulated by governmental regulatory agencies. In recent years, however, laboratories have increasingly been made the target of regulations formulated and enforced by various levels of government. Some of this regulatory attention is due to the scandals of the 1970s, which involved misreporting of data, falsification of records, and the design of experiments to satisfy a desired result (Broad and Wade 1982).

With the increased number and complexity of regulatory provisions generated in the past five to ten years, the laboratory manager or administrator must keep abreast of the regulations that govern his operations and see that all personnel within the laboratory are complying with them.

Since federal regulations in this area are complex and subject to change, the discussion that follows will be confined to an overview of regulatory provisions that are currently in effect. It is up to the reader to determine which specific regulations are pertinent to the operation of his laboratory and to obtain and review an up-to-date set of those regulations.

GOOD LABORATORY AND MANAGEMENT PRACTICES

Several regulatory bodies have published regulations pertaining to *good laboratory practices* (GLPs). These requirements specify the conditions under which studies must be conducted in order to be accredited by the regulatory agency. These conditions are reviewed whenever permits are submitted to the agency to market or distribute a new product.

Food and Drug Administration

The FDA is the responsible agency for enforcement of the Food, Drug, and Cosmetic Act, the Public Health Service Act, the Radiation Control for Health and Safety Act, and the Medical Device Amendment of 1976. These acts require that industry submit data to the FDA for the purpose of ensuring the safety of human and animal drugs, human biological drugs, medical devices and diagnostic products, food additives, color additives, and electronic products. To meet its consumer protection responsibility, the FDA requires that extensive animal and other types of testing be conducted (see appendix B). The requisite nonclinical tests are the responsibility of the manufacturer, who must demonstrate the safety of the product. The FDA's responsibilities include the following:

- Prescribing the type and extent of testing that the agency believes will provide the necessary data for FDA decision making
- Reviewing submitted data to determine whether safety standards have been met
- Carrying out independent studies in some cases to evaluate the safety of the product and to determine the validity of the test systems

The procedures are designed to answer the following types of questions:

- Can the product be safely tested in clinical studies with humans and animals?
- Are adverse effects likely?
- What is the toxicity profile?
- Are there potential teratogenic, carcinogenic, or other degenerative or dysfunctional effects?
- What is the no-effect dose in the test system?
- What level of use in foods can be safely supported?

Because of the importance of nonclinical laboratory studies to decisions regarding these questions, all studies must be conducted according to scientifically sound protocols and with detailed attention to quality control. To assure the quality and integrity of the nonclinical laboratory studies made to support applications for research or marketing permits for regulated products falling within the scope of the FDA's responsibilities, studies must be conducted in accord with the "Good Laboratory Practice (GLP) Regulations for Nonclinical Laboratory Studies" (21 CFR 58).

Environmental Protection Agency

The EPA issued a set of regulations in 1983 governing the conduct of studies to be used for registration of compounds under the Federal Insecticide, Fungicide, and Rodenticide Act (FIFRA) (40 CFR 162) and the Toxic Substances Control Act (TSCA) (40 CFR 702-75). These regulations were based on the EPA's belief that some studies submitted in support of regulated products were not conducted in accordance with accepted practice.

The GLP regulations for pesticides are divided into subparts A through H. The format for these regulations is similar to that of the FDA. However, the EPA regulations specify effects of noncompliance including legal action. Each of the subparts addresses different aspects of conducting this type of study. An outline of the points covered by these regulations is shown in table 10-1.

At this writing, the EPA and the FDA have entered into an interagency agreement to conduct laboratory inspections and study audits for

Table 10-1. Outline of Major Points Covered by EPA Regulations: Good Laboratory Practices

A. General Provisions
  Scope
  Definitions
  Applicability
  Compliance
  Inspections
  Effects of noncompliance
B. Organization and Personnel
  Personnel training
  Personnel qualifications
  Management
  Study director
  Quality assurance
C. Facilities
  Animal care
  Laboratories
  Specimen and data storage
  Management
D. Equipment
  Design
  Maintenance
  Calibration
	E. Testing Facilities Operations
	  Standard operating procedures
	    (SOPs)
	  Reagents and solutions
	  Animal care
	F. Test and Control Substances
	  Characterization
	  Handling
	  Mixtures
	G. Protocol for Conduct of a Study
	H. "Reserved"
	I. "Reserved"
	J. Records and Reports
	  Reporting
	  Storage
	  Retrieval
	  Retention
	K. "Reserved"

health effects testing. Additionally, the EPA has begun training its inspectors to audit both toxicological and environmental effects data and to conduct follow-up investigations.

A proposed GLP program for your laboratory to comply with the EPA's regulations should address, at a minimum, the following points.

- Organizational charts and functions
- Personnel qualifications
- Personnel functions
- Personnel training
- Floor plan of your facility
- Equipment list: calibration procedures; maintenance schedules
- Standard operating procedures
- Checklist for conduct of a study
- Quality assurance unit

Since these regulations are relatively new, you will have to pay close attention to changes as published in the *Federal Register* and modify your GLP program accordingly.

Occupational Safety and Health Administration

OSHA was created by the Occupational Safety and Health Act of 1970. While the Occupational Safety and Health Administration does not specifically regulate laboratory activities, the publication of a standard or guideline specific to laboratory operations is being considered. As a manager or administrator of a laboratory, you should be alert to changes in regulations as they are published in the *Federal Register*.

Even though your laboratory operations are not specifically regulated, you are subject to the regulations for general industry and could be visited by OSHA inspectors on a routine basis or as the result of a complaint filed by one of your employees. If violations of health and safety standards are found, you could be subject to fines of $1,000 for each violation.

Since the *General Industrial Safety and Health Standards* (OSHA 1978) comprise a fairly thick document with over a half million words, no attempt will be made here to summarize these regulations. (The standards can be obtained from the Government Printing Office in Washington, D.C.) Some general areas of which you should be aware in planning your overall safety program include the following:

- General laboratory design—walking, working surfaces and exits
- Protection from hazardous materials

- Exposure of employees to hazardous fumes or dusts
- Protective equipment
- Working environment—sanitary conditions
- Proper signs and tags for potential hazards
- Electrical protection
- Fire protection
- Medical and first aid facilities
- Noise exposure
- Training
- Record keeping

OSHA does require that each employer keep records regarding injuries and illnesses suffered by their employees. If an accident results in a fatality or the hospitalization of five or more employees, a report must be made to the nearest OSHA area director within forty-eight hours after it occurs. OSHA also requires that

- A report be obtained on every injury that requires medical treatment other than standard first aid
- Each injury be recorded on OSHA Form No. 200
- A supplementary record of occupational injuries or illness for recordable cases be prepared either on OSHA Form 101 or that workman's compensation reports having the same information be prepared
- An annual summary (OSHA Form No. 200) be prepared and posted no later than February 1 and be kept posted until March 1 of each year
- Records be retained for at least five years

National Institute for Occupational Safety and Health

NIOSH was created by the Occupational Safety and Health Act of 1970 to investigate safety and health hazards in the workplace. After identifying and evaluating a potential hazard, NIOSH recommends procedures for prevention and control. However, NIOSH has no authority to set and enforce standards as these activities are the responsibility of OSHA.

Standards for employee exposure limits to hazardous materials are recommended by NIOSH for inclusion in the OSHA regulations. NIOSH also performs the following functions:

- Testing and certification of personal protective equipment
- Education of personnel responsible for occupational health and safety

- Development of sampling and measurement methods to evaluate workplace hazards
- Publication of good practices for handling of specific materials in specific industries (tanneries, pesticide formulators, users of carcinogens, and so on)

DISPOSAL OF HAZARDOUS WASTES

In 1976, Congress passed the Resource Conservation and Recovery Act (RCRA), which controls hazardous wastes from generation to ultimate disposal. RCRA is administered and enforced by the Environmental Protection Agency. While the major regulatory effect of RCRA is on large manufacturing plants, laboratories can also come under its provisions.

Laboratories generate a wide variety of wastes, some of which may be classified as hazardous under RCRA. The volume of wastes generated by laboratories is generally smaller than that generated by a production facility. RCRA attempts to recognize this fact by allowing a small generator exclusion. Those locations generating a small volume of waste, currently less than 100 kilograms per month of hazardous wastes and one kilogram per month of acutely hazardous wastes, are not covered by the provisions of RCRA. These standards are undergoing review, however, and the EPA may soon issue guidelines for small generators.

More recently, the EPA has passed the responsibility of enforcing waste disposal regulations to various states whose waste disposal regulations are at least as stringent as the federal ones published as Parts 260-65 of Title 40 of the Code of Federal Regulations (40 CFR 260-65). Since state regulations may be more stringent than those published in 40 CFR 260-65, it is extremely important to become familiar with the applicable state regulations as well as the federal ones.

The hazardous waste regulations require that the generator of wastes be responsible for their safe management. Federal regulations recognize three main areas of participation in the management of hazardous wastes. These are the generator of the wastes, the transporter of the wastes, and the operator of the treatment storage and disposal facility (TSDF). Regulations allow for on-site storage of nonregulated quantities for indefinite periods. Regulated quantities (at least 100 kilograms of hazardous wastes and one kilogram of acutely hazardous wastes) may be stored for up to ninety days on site. If your laboratory were to generate one kilogram of a waste classified as acutely hazardous and stored this material for a period exceeding ninety days, you would be required to have both a generator permit and a TSDF permit. Any shipments of hazardous wastes from your laboratory to an approved TSDF are covered by Department of Transportation (DOT) regulations regarding packaging, labeling and identification of the hazard on the transport vehicle.

A hazardous waste is defined as a solid waste that exhibits toxic or hazardous characteristics. Specific hazardous wastes and acutely hazardous wastes are listed in 40 CFR 261. Even though a waste is not specifically listed, it may be classified as a hazardous waste if it meets one of the following criteria:

A. Ignitability
   1. Is a liquid with a flash point less than 60°C (140°F), determined by Pensky-Martens Closed Cup Tester or equivalent method
   2. Is not a liquid but liable to cause fires through friction, absorption of moisture, spontaneous chemical changes, retained heat from manufacturing or processing; or when ignited, burns vigorously and persistently, creating a management hazard
   3. Is an ignitable compressed gas as defined by DOT (flammable gas)
   4. Is an oxidizer as defined by DOT

B. Corrosivity
   1. Is aqueous and has a pH of less than or equal to 2 or greater than or equal to 12.5
   2. Corrodes steel (SAE 1020) at a rate greater than 0.250 inch per year at a test temperature of 130°F
   3. Is corrosive according to DOT

C. Reactivity
   1. Is normally unstable and readily undergoes violent chemical change without detonating; reacts violently with water; forms potentially explosive mixtures with water; or generates toxic gases, vapors, or fumes when mixed with water; or is a cyanide- or sulfide-bearing waste that can generate toxic gases, vapors or fumes when exposed to mild acidic or basic conditions
   2. Is capable of detonation or explosive reaction but requires a strong initiating source or must be heated under confinement before initiation can take place, or reacts explosively with water

3. Is readily capable of detonating or explosive decomposition or reaction at normal temperatures and pressures
4. Is a forbidden class A or class B explosive according to DOT (These wastes include pyrophoric substances, autopolymerizable material, and oxidizing agents.)

Two other criteria used to classify a waste are EP (extraction procedure) Toxicity (table 10-2) and toxic wastes (tables 10-3 and 10-4). Toxic wastes are divided into two groups, toxic and acute hazardous wastes. Toxic wastes exceeding quantities greater than 100 kilograms per month and acute hazardous wastes exceeding quantities greater than one kilogram per month are regulated.

General Responsibilities

All generators of hazardous wastes have certain responsibilities to carry out under either federal or state regulations. These include

- Determination of whether the waste is hazardous
- Notification of appropriate authorities of hazardous waste activities
- Maintenance of accurate records of wastes activities and filing of required reports

Table 10-2. Examples of EP Toxic Waste[a]

Contaminant	Extract Level (mg/l)
Arsenic	5.0
Barium	100.0
Cadmium	1.0
Chromium	5.0
Lead	5.0
Mercury	0.2
Selenium	1.0
Silver	5.0
Endrin	0.02
Lindane	0.4
Methoxychlor	10.0
Toxaphene	0.5
2-4-D	10.0
2,4,5-T	1.0

[a] A waste is classified EP toxic if the extract obtained by the EPA extraction procedure of a representative sample of the waste has concentrations of a contaminant that exceeds any of the values listed.

- Compliance with storage requirements
- Proper preparation of wastes for transport to the TSDF

Even though a generator may be exempt from the regulations due to a small-quantity exclusion, he may have a difficult time finding a TSDF facility that will take his wastes unless he has a generator permit and has made a detailed analysis of the waste shipment.

Costs for disposal of laboratory wastes are sky-rocketing at most TSDFs because of the complexity of handling, the record-keeping requirements, and the volume of wastes being generated. Additionally, violation of the regulations could be extremely expensive with fines ranging up to $25,000 a day per violation.

A generator must also meet the storage requirements of the regulations if wastes are accumulated for shipment to a TSDF. These include placing the wastes in containers in good condition in a well-maintained

Table 10-3. Examples of Toxic Wastes (designation T)[a]

Acetaldehyde
Acetone
Benzal chloride
Benzotrichloride
Ethyl acetate
Ethyl acrylate
Ethyl ether

[a]Complete listings may be found in 262.33 paragraph (f) of RCRA regulation, 1980.

Table 10-4. Examples of Acute Hazardous Wastes (designation H)[a]

Aldrin
2,4-D
Dieldrin
Endosulfan
Methyl parathion
Carbon disulfide
Chloroacetaldehyde
Methyl isocyanate
Thiosemicarbazide

[a]Complete listings may be found in 262.33 paragraph (e) of RCRA regulations, 1980.

area, marking the container with the date accumulation started and properly labeling it as containing a hazardous waste.

One provision of the RCRA regulations requires that any shipment of hazardous wastes off site be accompanied by a manifest tracking it from the generator to the disposal site. Copies of the manifest need to be available for each handler—for the generator, the transporter or transporters, and the disposal facility. Copies of the manifest must also be filed with the responsible regulatory agency.

Because of the complexity of the regulations, consideration for environmental pollution, and the costs of disposal, most laboratories will exercise waste management options to minimize the amount of hazardous wastes generated. One of the most effective options is to limit the amount of materials brought into the laboratory to the quantities required to do the work. Another method is to neutralize or chemically alter the by-products of laboratory reactions to a form not classified as hazardous. A third option is to recover or regenerate wastes such as solvents so that the products are reusable. Finally, laboratory wastes may be changed to nonhazardous products through incineration (care should be taken to determine that any laboratory incinerator meets state air-quality discharge regulations).

POLLUTION CONTROL

In addition to the hazardous wastes generated by your laboratory, you must be aware of the rules and regulations of various regulatory agencies governing the quality of the air that is exhausted from your laboratory and the material that is discharged to either sanitary or storm sewers. These agencies may include

- State air quality agencies
- Sanitary sewerage districts
- City environmental quality agencies

The laboratory manager will need to obtain copies of the regulatory standards that are applicable to his location.

SHIPPING HAZARDOUS MATERIALS

Shipments of hazardous materials are regulated by the Department of Transportation under Title 49, Code of Federal Regulations. These regulations cover packaging, labeling, and shipment of materials ingredient derived from rDNA has been formally considered by the FDA, so no clear regulatory route has been determined.

Environmental Protection Agency. The EPA has asserted that it has regulatory authority over rDNA products, reasoning that recombinant DNA is a new or altered form of deoxyribonucleic acid. As a new chemical substance, rDNA, along with the organisms carrying it, is presumed to fall within EPA jurisdiction under the Toxic Substances Control Act (TSCA). In addition, the EPA has established a precedent by taking responsibility in areas where other agencies have been unable to establish regulations. The EPA's general mandate to protect the environment will probably permit the agency to regulate products, by-products, and wastes from rDNA.

As of this writing, no legislation or guidelines have been published to explain exactly how the EPA will regulate all of the various kinds of products that might eventually be derived by rDNA technology, except for pesticides. Before these are established, it is highly probable that the EPA's interpretation of TSCA will undergo legislative and judicial review.

U.S. Department of Agriculture. The USDA regulates vaccines, serums, and related products used to treat domestic animals under the Virus, Serum and Toxic Act. Providing for licensing of products and factories, this act gives the USDA wide authority to require just about any information it needs to judge product purity, safety, and efficacy. Indications are that it will probably be difficult to obtain approval for any live, genetically engineered organism to be used in vaccines or other biologicals. Other animal drugs, such as growth hormones, would be covered under the FDA's Bureau of Veterinary Medicine and would require a "New Animal Drug Application" (NADA).

Occupational Safety and Health Administration. OSHA is evaluating and developing a policy on rDNA under the Occupational Safety and Health Act, because of its concern for the protection of workers from existing or potential hazards that might be specific to this technology *(Federal Register)*. The National Institute for Occupational Safety and Health (NIOSH) is assessing potential hazards associated with industrial applications of rDNA technology. The investigation includes on-site inspections of industrial plants to review physical containment design, engineering, controls, work practices, validation procedures, and emergency procedures along with monitoring for potential personnel exposure. OSHA's authority to regulate rDNA comes from the act's mandate to "assure safe and healthful working conditions." It appears that any alleged biohazard associated with rDNA would be covered by the terms of this act.

States and Municipalities

Regulations have been passed by various states and municipalities that tend to fill voids left by federal regulations. Various cities as well as New York state and California have enacted legislation placing added restrictions on rDNA research activities.

Current Guidelines

The regulatory picture for rDNA and the products derived from the technology is complex and likely to become even more so in the future. For now, the document that provides the most positive and concise rules to follow is the NIH guidelines (1980). The guidelines are primarily oriented toward research-scale activities, but they also include containment requirements for large-scale and production work (defined as greater than ten liters). The principal purpose of the guidelines is to specify practices for facility construction and handling methods for recombinantly derived molecules, organisms, and viruses.

Classes of Experiments. All research and production involving rDNA is divided into classifications according to the degree of potential hazard based on the host-vector system, the quantity of viable material, or the potential for release of viable material from the process. Classifications based on amount and level of administration required are identified below, along with their requirements for approval and notification:

   A. Experiments that require RAC (rDNA Advisory Committee) approval prior to initiation
      1. Use of rDNA coding for toxic molecules (LD_{50} less than 100 nanograms/kg — for example, botulism, tetanus, or diphtheria)
      2. The deliberate release of rDNA organisms into the environment
      3. Transfer of resistance factors into a host that does not naturally acquire that resistance factor
   B. Experiments that require IBC (institutional biosafety committee) approval prior to initiation
      1. Use of class 2 through 5 human or animal pathogens
      2. Transfer of DNA from human or animal pathogen to nonpathogenic prokaryotic or eukaryotic host-vector system

3. Use of infectious animal or plant viruses or defective viruses plus helper viruses in tissue culture
4. Use of whole animals or plants
5. Use of volumes of culture greater than ten liters

C. Experiments that require IBC notification concurrently with initiation of the experiment
   1. Use of components that are all derived from nonexempt, nonpathogenic organisms
   2. Use of less than 2/3 of a viral genome

D. Experiments that are exempt from prior approval, but not from the guidelines
   1. Use of only DNA segments from a single, nonchromosomal, or viral DNA source
   2. Use of DNA from and within a single host or a closely related strain of the same species (including eukaryotic)
   3. Use of different species that exchange DNA by known processes (A list of these is included in the guidelines.)

E. Experiments approved on a case-by-case basis by the Director of NIH after RAC review

Roles and Responsibilities. Specific guidelines have been set forth for managing and carrying out rDNA activities. The key roles have been identified below, along with some of their major functions.

A. Institutional responsibilities include
   1. Establishing procedures and policies in accordance with the NIH guidelines
   2. Establishing the institutional biosafety committee (IBC)
   3. Appointing a biological safety officer if research is conducted under P3 or P4 conditions
   4. Requiring investigators to comply with the guidelines
   5. Ensuring appropriate training of all personnel
   6. Training IBC members regarding the guidelines and laboratory safety
   7. Determining the necessity for health surveillance and, if necessary, instituting a program in accordance with the *Laboratory Safety Monograph* issued by NIH (1979)
   8. Reporting research-related illnesses or accidents

B. Membership and functions of the IBC include
   1. Maintaining active membership of at least five members, with at least two from outside the organization

2. Reviewing research to determine compliance with NIH guidelines including containment levels, training, and expertise of personnel (The IBC may set or lower certain containment levels as specified and must adopt emergency plans for accidents and spills.)
C. Responsibilities of the biological safety officer include
   1. Providing technical expertise and guidance to the IBC and the scientific investigators
D. Responsibilities of a principal investigator (PI) include
   1. Ensuring compliance with the guidelines
   2. Reporting to the IBC, training personnel, and adhering to emergency plans
   3. Supervising the safety performance of staff, providing for the correction of work errors, and ensuring the integrity of physical containment facilities

Large-Scale Work. If rDNA work is to be carried out at volumes greater than ten liters, the institution must carry out some additional responsibilities:

- A biological safety officer must be appointed.
- If research is carried out under P3 or P4 conditions, a health surveillance program must be established and maintained.
- Good laboratory practices must be followed, with additional requirements depending on the classification of the research.
  - All viable material must be handled in a closed system and inactivated prior to removal.
  - Sampling and inoculation of a closed system must be carried out to minimize the release of aerosols.
  - Off-gases must be properly handled via HEPA filters or incineration.
  - Closed systems must be sterilized prior to opening for maintenance.
  - Emergency plans must be established for handling spills or accidents.

Laboratory Procedures. Sections on the P1 and P1-LS physical containment levels briefly summarize certain requirements for good, safe laboratory practices. Additional procedures for other laboratory and large-scale containment levels are specified. As stated in the guidelines, the first principal of containment of rDNA molecules and organisms is the strict adherence to good microbiological practices (NIH 1984).

Certain basic procedures are essential to this level of containment.

While most are elementary (good laboratory practices and common sense), some areas deserve particular attention and are mandatory for rDNA research.

- Keep laboratory doors closed.
- Decontaminate work surfaces daily.
- Decontaminate waste, glassware, and laboratory equipment prior to washing and reuse.
- **DO NOT** pipette by mouth! Always use a mechanical pipetting device.
- Do not eat, drink, or smoke in the laboratory.
- Wash hands after handling rDNA materials and before leaving the laboratory.
- Conduct all procedures so as to minimize aerosols. (Equipment that creates large quantities of aerosols includes blenders, sonicators, and stirrers. Special care and/or containment is called for when performing procedures with equipment of this type.)
- Transport materials to be decontaminated outside the laboratory in approved, leak-proof containers.
- Use lab coats, gloves, and safety glasses or goggles. Remove protective clothing prior to leaving the laboratory.
- Properly utilize biological safety hoods.

Two major sources of problems in containment areas deserve attention. They are the use of the laminar flow hood and the use of equipment that produces aerosols. Laminar flow biological safety hoods are designed to be used with the sash in a particular position. When used with the sash open, efficiency is diminished and the distribution of viable organisms outside the cabinet may even be worse than if the work were done without the hood.

Aerosols can be produced by various procedures and equipment. Many experienced laboratory people do not realize that a simple action such as blowing the last drop out of a pipette can cause aerosols to be dispersed fifteen feet or more.

CONTROLLED SUBSTANCES

The Federal Food, Drug, and Cosmetic Act provides authority to the Drug Enforcement Agency (DEA) of the Department of Justice to control the use of and protect the public from abuse of certain substances deemed to have significant physical and/or psychological effects on people. With allowance for specific types of exemptions, five schedules of substances are provided in Title 21 of the Code of Federal Regulations. These lists are updated annually by the DEA.

The laboratory manager will occasionally find it necessary to work with these regulations since many of the chemicals have a valid use in clinical and biological chemistry applications. Detailed records and reporting are required from laboratories that make use of or store any of the items on the schedules. Records must be kept on each use of each substance, the quantity used, residual materials, by-products, and disposal. Waste, product, and by-products must then be controlled for disposal or sale under specified rules. Reports are provided to the DEA annually indicating the substance, quantity purchased, quantity on hand, and use or disposal of all other amounts. The laboratory manager must assign a very reliable individual to ensure that proper records are maintained. It must also be noted that there are specific labeling requirements for all vessels that contain one of these substances or by-products. The label must include, in a prominent way, a statement such as "For Industrial Use Only" or "Diagnostic Reagents—For Professional Use Only." The schedule or exact ingredients is not required.

Drug Enforcement Agency

The five schedules are described in brief below:

- *Schedule I* drugs or other substances
  - Have a high potential for abuse
  - Have no currently accepted medical use for treatment in the United States
  - Lack accepted safety data for use under medical supervision
- *Schedule II* drugs or other substances
  - Have a high potential for abuse
  - Have a currently accepted medical use for treatment in the United States or a currently accepted medical use with severe restrictions
  - May lead to severe psychological or physical dependence or high psychological dependence if abused
- *Schedule III* drugs or other substances
  - Have a potential for abuse less than those in schedules I and II
  - Have a currently accepted medical use for treatment in the United States
  - May lead to moderate or low physical dependence or high psychological dependence if abused

- *Schedule IV* drugs or other substances
  - Have a low potential for abuse relative to those in schedule III
  - Have a currently accepted medical use for treatment in the United States
  - May lead to limited physical dependence or psychological dependence relative to those in schedule III if abused
- *Schedule V* drugs or other substances
  - Have a low potential for abuse relative to those in schedule IV
  - Have a currently accepted medical use for treatment in the United States
  - May lead to limited physical dependence or psychological dependence relative to those in schedule IV if abused

The laboratory manager must obey the regulations and keep good records because strict criminal penalties are imposed for noncompliance. He must be sure employees in his laboratory are fully aware of the precautions and the hazards.

THE FDA AND FOOD ADDITIVE OR GRAS APPROVAL

This section is intended to provide information about the complexity of regulations and interactions with the various agencies for the laboratory manager who is working with products that will be used in the manufacture or preparation of foods. Familiarity with regulations will aid the development of new bioproducts and help speed the safety evaluation and regulatory approval process. Information is included on the data required to determine the proper testing category and the material required for inclusion in petitions for GRAS ("generally recognized as safe") or food additive approval.

Products of this category are ultimately destined for use in foods or in food ingredients. In order for the FDA to permit inclusion, the product must be an approved food additive, GRAS, or a prior-sanctioned substance (in common use prior to 1958). Essentially the same information is required for all petitions, but a food additive cannot be marketed until after it receives final approval from the FDA, whereas a GRAS substance can be marketed immediately after submission of the petition. The choice of approval category depends on the nature of the product. These products may be derived chemically or found naturally and processed, or be a bioproduct derived by fermentation techniques. Those that are chemically based or natural require the same approvals as bioproducts. For bioproducts, information is required concerning the source organisms.

Microorganisms

Microorganisms are not themselves approved for use in foods or in the production of food ingredients. It is the products derived from specific organisms that are approved, with the organism listed as the source. Products from only a few microorganisms are listed by the FDA. Therefore, it is usually necessary to seek some level of regulatory approval for most new products derived from an organism. If a product from a particular organism has been previously approved, it does not automatically mean that different products from other strains of the same species would not require further testing or even separate regulatory approval. Therefore, it is important that the proper information be supplied in order to evaluate the approval requirements of each product.

Taxonomic characterization of genus, species, subspecies, and strain designation needs to be provided (Krieg 1984). This work may be carried out in your own laboratory using recognized techniques and classification schemes, if proper facilities and qualified personnel are available. It should be contracted to an outside laboratory in cases where specialized equipment or procedures are necessary or when the opinion of a recognized expert is needed.

The pedigree should be as complete as possible. Information regarding the origin of the organism, in-house handling, and storage should be included. In the case of a recombinant organism, history of the origin and complete characterization of the cloned DNA, the vector, and the host systems will probably be required by the FDA. (The regulation of recombinant DNA organisms remains subject to change by the various regulatory agencies.)

Literature searches for information regarding the history of the organism's use in food, its presence as inhibitor or contaminant in food and soil, and its history of association with disease or toxin production are necessary. Toxin producers in the species should be identified, together with the presence and mode of pathogenesis in the same and related species.

Safety Evaluation. A main aspect of the safety evaluation of new bioproducts includes the testing of the organism for pathogenicity, toxicity, and the production of antibiotics. Testing for pathogenicity and toxicity can be carried out by a contract toxicology laboratory. The methodology employed may vary depending on the suspected route of entry and mode of pathogenesis for the organism in question. Likewise, toxicity potential may be evaluated by one or more methods in one or more animal species.

It may also be necessary or advisable to test for toxins by analytical chemistry techniques. These tests can be carried out in house or by a contract laboratory. Specific protocols must be decided upon on a case-by-case basis. Testing of the organism for the production of antibiotic substances may be carried out in house if definitive information indicating nonproduction is not available.

Components

Components are substances added to or used in the preparation of food additives and must be considered in the safety evaluation of the final product. These could include such minor components as defoaming agents, pH control agents, flocculating agents, and preservatives. Components used in the preparation, purification, dilution, standardization, or preservation of the final product must be demonstrated to be safe for the specified use. Data to establish the safety or food approval of components is required in order to receive product approval.

Information regarding the history of use of a substance in food or food processing is helpful in evaluating its safety. If use prior to 1958 can be documented, its regulatory status might be clearer since such substances may be "prior sanctioned." As much information as possible should be supplied so that the current regulatory status can be determined. Usually it is helpful to have information on the chemical name, Chemical Abstract Services (CAS) registry number, and common names in order to determine whether the substance is food-approved or not. If a component is not well known, previously food-approved, or GRAS, it is necessary to include as much information as possible regarding its safety and suitability for the intended use.

In evaluating the safety of a component or processing aid, a prime consideration is the determination of the residual level in the final food to which the consumer might be exposed. Usually, a worst-case risk analysis is carried out. In order to calculate this, complete information needs to be provided about the chemistry, use levels, and fate of the component in the final product.

The Final Product

The primary consideration in evaluating the safety of a product is the product itself. There are several aspects to this evaluation to which the FDA has applied guidelines, particularly those set forth in *Toxicological Principles of the Safety Assessment of Direct Food Additives and Color Additives Used in Food* (also known as the "Red Book") (FDA 1982). This document provides guidelines for categorization and testing of various substances and aids in deciding upon a testing protocol sufficient to demonstrate the safety of a particular additive.

Exposure Level. One of the first exercises to be carried out on a proposed new food additive is the estimation of the level to which the consumer will be exposed. As with components, to accomplish this it is necessary to obtain sufficient information regarding the chemical and physical aspects of the use of the product to enable the calculation of exposure levels on a worst-case basis. This information should include a description of the proper method of using the product and any additional information that would aid in the calculation.

Safety Studies. The Red Book sets forth specific guidelines for testing a food additive based on several factors. Among these are the level of consumption as determined above and the chemical category into which the substance falls. It is important, therefore, to know the chemical structure or class of compounds to which the substance belongs. Based on these factors, the extent to which testing is necessary varies widely from those of the least concern requiring only a twenty-eight-day feeding study and short-term tests for carcinogenic potential to the highest level requiring extended long-term animal studies.

Space does not permit a description of the details of structure categorization, but this information is readily available (FDA 1983). Once the structure category and exposure level have been determined, it is possible to decide the level of concern and the level of testing to which the substance must be subjected. As an example, below is a brief description of the tests required for Concern Level III compounds. (Test requirements for Concern Levels I and II can be found in the Red Book.)

1. Carcinogenicity studies in two rodent species
2. A chronic feeding study of at least one year in duration in a rodent species (Under most circumstances this study is added to one of the carcinogenicity studies and performed as a combined test.)
3. Long-term (at least one year in duration) feeding study in a nonrodent species
4. Multigeneration reproduction study (minimum of two generations) with a teratology phase in a rodent species
5. Short-term tests for carcinogenic potential that can be used for

determining priority for conduct of lifetime carcinogenicity bioassays, and that may assist in the evaluation of results from such bioassays.

Short-term tests for carcinogenic potential could include the Ames test, the mouse lymphoma test, and the unscheduled DNA test, as well as other tests for genotoxicity. These would be selected based on appropriateness for the particular compound or substance under study. For instance, the FDA has agreed that the Ames test would probably not be appropriate for testing of an enzyme preparation because of the high probability that histidine would be present.

Analytical Methods. Analytical methods for the final product and many, if not all, of the components of the product are essential. These should be developed in order to support the safety claims made for the product. The sensitivity of each method must be appropriate to detect levels of the substance at levels equal to or below the claimed residual levels found in the final product.

The Petition

The final step in the process of obtaining GRAS affirmation of approval as a food additive is preparation of the petition for submission to the FDA. Basically, this consists of a compilation of the data that has been discussed in this section. The completed petition would contain information encompassing the following:

- Name, chemical identity, and properties
- Amount, purpose, and tolerances
- Data establishing efficacy of the intended use
- Method of use
- Full safety data for additives and components
- Environmental impact of the manufacturing process

This large volume of data must be submitted to the FDA. Review and approval will take time and probably require much communication with the assigned FDA investigator.

FDA DRUG AND DEVICE APPROVALS

The Food, Drug, and Cosmetic Act and subsequent amendments provide the Food and Drug Administration (FDA) the authority to ensure the efficacy of claims made concerning ethical drugs, over-the-counter (OTC) drugs, antibiotics, and medical devices. These include articles for which a claim is made intending a cure, treatment, prevention, or diagnosis of a condition, even if it is not claimed to be a medical product. A major portion of the legislation is concerned with adulteration and misbranding. It also includes the specification of requirements for new drugs, OTC drugs, antibiotics, biologicals, and controlled substances. The primary function of the regulation for new drugs is the satisfaction that it is either (1) generally recognized as safe and effective for the use defined, or (2) proven to be safe and effective for its intended use through the pre-market approval process. This book cannot possibly cover all the volumes of legislation that exist, but is intended to aid the laboratory manager in understanding some of the complexities involved in the process of obtaining drug or device approval. For futher information on the regulatory aspects of development and production of drugs and devices, consult the regulatory affairs personnel in your organization, your reference library, or your legal staff. The detailed requirements for drugs and devices are found in section 21, CFR. It is here that the laboratory manager or regulatory affairs specialist will be able to find the details of good manufacturing practices (GMP) as currently defined for drugs and devices. The major aspects of product design, development, and manufacture that are covered and for which the laboratory manager may have some responsibility include those that follow. The emphasis here is not on providing answers, but on helping the manager identify questions that need to be asked and areas that should be researched before proceeding too far into a project.

General Provisions

The basic requirements for a manufacturer in ensuring quality and efficacy of the products are outlined in the following paragraphs.

Organization and Personnel. This area defines the required roles to meet GMP and the responsibilities of the personnel involved in manufacture and quality control, as well as the appropriate training for those individuals.

Buildings, Facilities, and Equipment. This section outlines the control and engineering requirements that are necessary for a facility to meet FDA standards in terms of cleanliness, air quality, process utility quality, validation of equipment, and proper control and maintenance of all processes.

Product Testing and Pre-Market Approval. This specifies the laboratory, toxicity, and animal tests that must be carried out for a specific product to prove efficacy and safety.

Component Control. This part of the CFR sets forth the holding requirements for raw materials, the testing necessary for qualification of materials, the quarantine requirements to ensure efficacy, and the release approval guidelines.

Production and Process Controls. In this section, the conditions that must exist in production areas are described, as well as the controls required on all equipment, including printed data, maintenance, and monitoring.

Packaging and Labeling Control. This section deals with safeguards required for packaging and the information necessary for labeling.

Holding and Distribution. The following questions are taken up under this heading: How long can products be held? To what extent must the distribution and location of units and components be known? To what level must a business be able to trace a small vial of an ethical drug and where it has been distributed? What tests must be performed before a finished product can be released from quarantine?

Device Installation and Evaluation. Guidelines are given concerning information that is required to be passed on to the user concerning installation, conditions for installing the equipment, and the evaluation required to determine that a product lives up to its claims.

Laboratory Controls. The conditions that must exist in the laboratory to meet GMP and GLP standards are outlined.

Records and Reports. Requirements are specified concerning the recording of data, the number of copies of all records that must be maintained, and for what periods, and kind of complaint files that should be kept.

Returned Products. The obligation of the company to the purchaser is described.

Ensuring Quality

The critical reason that the laboratory manager, and all others within the company that have any responsibility for meeting GMP and product efficacy, must be cognizant of these requirements and follow them is product liability. The repercussions to a company for failure to ensure product safety and the safety of the people using the product may be catastrophic. The company must have on file evidence that ensures that the product has been manufactured reliably. All equipment must be certified and all processes validated to show that the intended performance characteristics are met through each stage of the process. The FDA requires a validation program that "provides the high degree of assurance that a specific process will consistently produce a product meeting predetermined specifications in quality." The Pharmaceutical Manufacturers Association (PMA) has proposed their own definition of validation to be "documentation that ensures that a process performs as it is claimed to." Quality cannot be tested into a product. It must be designed and built into the product. By the small number of samples taken in validation and quality assurance, testing cannot in itself ensure product quality, and therefore extensive documentation of the processes must be maintained to ensure that product quality is achieved. This is definitely not a one-man job. In fact, the responsibility does not belong to a single department. It takes a team effort and many departments to design and produce a high-quality product. The departments and activities that may contribute to this effort are listed in table 10-5.

The major reasons for a company to go to such lengths are to (1) provide a product that is safe and efficacious and (2) prevent complaints, illnesses, or damages that could result from use of the product. The purpose of the extensive documentation, besides serving as an inspection medium for the FDA, is to protect the company by providing a defensible position in the event of injury or illness. A side benefit of a thorough, sound validation program is that it helps to prevent or lessen the number of rejects, reworks, or possible recalls, thus saving the company money.

Table 10-5. Departmental Responsibilities for Validation and Documentation of Product Quality Assurance

Department	Activities
Research and Development	All the necessary scientific disciplines
Toxicology	Maintenance of animal colonies and toxicology testing
Regulatory Affairs	Interfacing with all necessary regulatory and legislative agencies
Engineering	Sterilization techniques, process engineering, packaging, and equipment development
Validation	Documentation development and test certification
Quality Assurance	Analytical services, stability testing, materials clearance, and documentation
Manufacturing	Documentation, GMP monitoring, and validation testing
Marketing	Product management and production planning
GMP Compliance	A group outside manufacturing acting as an internal watchdog and monitoring regulatory changes

EQUAL EMPLOYMENT OPPORTUNITY AND AFFIRMATIVE ACTION

The laboratory manager must be concerned with and knowledgeable about Equal Employment Opportunity (EEO) and Affirmative Action (AA). The reason is that almost every personnel action has a potential legal result. Discrimination suits can jeopardize contracts, result in litigation, or cause morale problems. If the supervisor is not knowledgeable about EEO, he exposes his company, agency, or institution to potential liability as would a person deliberately trying to circumvent AA efforts.

This may seem overstated, but it is a fact that many times the research manager sees EEO as something to be handled entirely by the personnel department, the local EEO officer, or lawyers. This section will illustrate why EEO and AA concern more than just hiring employees and are not simply a numbers game. EEO involves who you hire and also who you do not hire, who you fire, every transfer, training programs, promotions, pay scales, retirements, questions asked during each interview, disciplinary actions, employee evaluations, merit increases, and most other actions that affect personnel. These activities should sound familiar, as most of the items mentioned are the direct responsibility of the laboratory manager and other supervisory personnel.

EEO is not restricted to racial minorities, although that is the first thought of most supervisors. It includes employment opportunity and job protection for older workers, women, ethnic minorities, veterans, the handicapped, and as a result of some court decisions, white males. EEO applies to every personnel action with every employee.

The research manager should realize that it is much easier to file a complaint than it is to defend against one. The burden of proof does not fall on the person filing a complaint, but on the corporation, agency, or institution. These facts make it imperative that you and each of your supervisory people become sensitive to EEO/AA. Small EEO mistakes, the kind a well-intentioned, but uninformed supervisor might make, carry the risk of government intervention that can limit the effectiveness of the personnel department and disrupt the laboratory.

History

The EEO/AA picture is confusing because regulations overlap and are not uniformly interpreted by the courts. The basic intent of legislation in this area has been to allow individuals to file charges, obtain government intervention, and sue their employers or potential employers. The Civil Rights Act of 1964 was the most important of these laws. One section prohibits discrimination in employment based on race, color, religion, national origin, or sex. This act created the Equal Employment Opportunity Commission (EEOC), which was granted the power to investigate charges, make determinations of probable cause, seek conciliation, publicize problems, and issue interpretive guidelines (EEOC 1974). To increase the EEOC's effectiveness, it was later granted the power to file legal suits, either on behalf of a complainant or on its own. Regulations require that every employer with fifteen or more employees regularly file federal forms showing employment statistics by race, sex, and job category.

Other statutes extended the scope of the individual's rights against discrimination. They include the Fair Pay Act of 1963 (requiring equal pay for equal work), the Age Discrimination Act of 1967 (designed to protect a growing work force of age forty to sixty-five years), and the Vocational Rehabilitation Act (setting standards regarding handicapped workers).

In an executive order issued by President Johnson in 1965, Affirmative Action was first announced. It required government contractors to be "fair employers" and to establish procedures to ensure fairness. The request was not new, but the penalty for not following it was loss of present and future government contracts while the company or institution remained in noncompliance. Employers understood that E.O. 11246 and the new statutes called for color-blind employment practices, and they hurried to ensure compliance in order to protect contracts. Employers, regulators, and the judiciary were all unsure what AA meant or required, but everyone knew that immediate action was necessary. Time was required to clarify the regulations.

From Law to Practice

It has been said that it is not the legislature that decides what a law means, but the judiciary. Examining employment practices that the courts have found discriminatory is one way to understand the potential effects of judicial interpretations:

- The courts found that a job statement specifying "high school graduate" discriminates if the company could not easily show that a high school education was essential for success in the position.

- Advertisements indicating "under thirty-five," "young person," or "supplement your pension," have been judged discriminatory.
- Universities and other employers have been challenged to prove that having a Ph.D. or being published (which are proportionately rarer among minorities and, to some extent, women) are really essential to the job.
- Statistics showing that minorities or women receive less frequent promotions or smaller merit increases than white males have been accepted as evidence of systematic discrimination.
- Previous experience may not be a valid job requirement if the effect is to discriminate against minorities or women who historically did not have opportunities in a given profession.

Employment qualification tests generally fail to pass the rigorous validation criteria and testing is disappearing as a prerequisite to job placement.

Today's Status

Employees who thought that color-blind employment practices would be sufficient to satisfy the requirements proved to be wrong. The courts have declared that it is not just the intent of the practice, but its effectiveness that must be judged. If a personnel practice, although it appears fair on the surface, results in discriminatory actions or perpetuates past discrimination, it is judged to be in noncompliance. A company's policy of filling positions from employee referrals was judged discriminatory, because the work force remained nearly all white. Unequal treatment is not the main consideration in the courts. Now unequal impact must be considered.

This judicial interpretation is one factor that forced the trend towards emphasizing numbers and statistical analyses, in defense of the company and for investigations by outside agencies. Defending a charge of discrimination involves expensive research and legal costs. Keep in mind that a single discrimination charge opens the organization's records to investigating agencies.

The local, state, and federal agencies and other organizations that look for discriminatory actions expect companies to do more than practice indifferent equality. They expect active programs in recruiting, training, and education among the affected groups. The courts have stated: make the affected group "whole" and enable them to take their "rightful place."

Complaints have become easy to file, requiring only a telephone call, letter, or sworn complaint form in most cases. Employees have become more aware of the resources that have been made available to assist them. In order to aid the individual against the vast resources of the employer, the organization has been placed in the position of proving that the charge is not true, with the complainant having fewer responsibilities to demonstrate the charge is true. This kind of negative proof is difficult to provide.

The Company's Role

Your company, agency, or institution should have a written policy supporting the goals of EEO and AA (Calvert 1979). You should treat EEO as a necessary component of good management practice (Hayes 1980). Conflicts between employee and employer can and should be resolved by good two-way relations and communications. If a complaint is filed, other considerations are the added expenses, loss of productive manpower, and unfavorable publicity that results. Every employer and supervisor must make an effort in the area of equal employment opportunity to achieve two goals:

1. Treat all employees and candidates fairly and without regard to race, creed, sex, color, national origin, or other criteria not related to job qualifications.
2. Minimize the number of discrimination complaints filed against the organization and be prepared with adequate records and procedures to defend against all complaints that have no merit

The Supervisor's Role

A critical fact that the laboratory manager needs to remember when it comes to EEO/AA matters is that, to the people you supervise, you are the company. What they see you do will be looked on as the policy of the entire organization. Their opinions will largely be based on their dealings with you.

The following are some hints for handling a few of the "danger areas" where mistakes are most frequently made.

Placement. Review the job description and be sure that it accurately describes the nature and scope of the required work. Carefully consider whether the job requirements you have specified are appropriate and necessary. If a requirement cannot be easily proved essential to job success (if a job candidate who does not possess it can in fact perform the job tasks), then you have a potential complaint.

Lower the starting requirements in jobs where the tasks can be quickly learned. This is a good management practice since hiring an overqualified employee means paying for skills that are not used and may result in an unhappy employee.

Try to reassign work in your group, providing a path for advancement. Reorganization of the work may create an entry level or training position. Be aware that hiring women and minorities for these positions will not by itself meet affirmative action goals and may actually make things worse. It could improve the organization's overall statistics, but analyzing the breakdown by job category would tend to give an unfavorable picture of promotion practices. Try to promote from within whenever possible.

Interviewing. No questions have been expressly forbidden, but the EEOC has identified some that it looks upon with "extreme disfavor." These include

- Arrest and conviction records: (exceptions are made for certain recent and job-related convictions, such as a bank asking teller candidates about theft and embezzlement.)
- Garnishment records (Like arrests, these are discriminatory because some minority groups may have a higher percentage of occurrence than nonminorities.)
- Credit references
- Child-care, unless asked of both men and women (An example is the question, "If your child were sick, would you have to stay home?")
- Age
- Height and weight

In general, ask the same questions of men that you ask of women and the same questions of minorities as of nonminorities.

Some special cautions are in order when interviewing women. As some men still hold stereotypes, there is a frequent incidence of questions on plans for raising a family, alimony, freedom to travel, willingness of the husband to let her meet male business associates, and so on. Questions such as these are discriminatory. They are also unnecessary because they are not related to the job requirements. Even a caution that a particular job is too dangerous, too unpleasant, or in an all-male office where a woman is not welcome is unacceptable and has in the past been the basis for successful complaints.

Learn to control your own biases and remember that everyone has some. Remove your emotions from the process of selecting the best classified as hazardous by the DOT. CFR-49 states, "[I]t is the duty of each person who offers hazardous materials for transportation to instruct each of his officers, agents and employees having responsibility for preparing hazardous materials for shipment as to the applicable regulations." Regulations governing shipment of hazardous materials include

- Determination of the hazard class or classes
- Determination of proper shipping name and identification number
- Selection of the proper package
- Selection of the correct labels to be applied
- Marking of the package
- Preparation of shipping papers

For the most part, the regulations contained in CFR-49 were designed to cover shipment of commodity materials. This presents the laboratory manager with a dilemma regarding shipment of newly synthesized materials that have not been thoroughly tested to determine if they are in fact hazardous. If your laboratory does not have the capability to establish the toxicity of a material, you must rely upon an outside lab to conduct this type of test. The regulations do allow for shipping of samples for testing to determine potential hazards. However, it is up to you to assign the sample to hazard class, using your best knowledge of the material being shipped. Currently there are twenty-one different classifications of hazardous materials described in CFR-49.

The regulations cover shipment of materials by any means—motor freight, private vehicle, parcel services, air freight, and ships. Laboratories that are on a site divided by a public highway could be subject to DOT regulations for transfer of materials from one building to another building on the site if the highway had to be crossed. It is also important to advise your technical personnel of these regulations. In the past, many samples have been transported by technical or sales personnel from one location to another by placing the sample in a briefcase, suitcase, or pocket. Fines under these regulations are substantial, up to $10,000 per day for each day the violation exists or up to $25,000 and five years in jail for a willful violation.

International shipments are covered by similar regulations. However, these regulations may not be identical to those of the DOT. A good reference source covering the shipment of materials by air is *Dangerous Goods Regulations,* published by the International Air Transport Association (1985).

RECOMBINANT DNA

The regulation of recombinant DNA (rDNA) technology has become nearly as complex as the technology itself. What began as a fairly specific set of guidelines has drawn the attention of numerous agencies, each wanting to regulate the technology or the products derived from it. This section is intended to provide an overview of the regulatory status and a feel for the direction it may take in the future.

The National Institutes of Health

The National Institutes of Health (NIH) *Guidelines for Research Involving Recombinant DNA Molecules* (1980) is the primary reference. The guidelines are updated periodically and published in the *Federal Register*. They apply directly to federally funded institutions, grants, and programs, and noncompliance can result in suspension of funding. The guidelines also serve a quasi-regulatory function with respect to the biotechnology and genetic engineering companies involved in rDNA research. The policy of most companies, especially those belonging to organizations such as the Industrial Biotechnology Association and the Health Industry Manufacturers Association, is to abide by the guidelines voluntarily. Even though the guidelines do not serve a true regulatory function, they would probably serve as evidence of "reasonable care" if a company were involved in a liability action resulting from an accident or injury involving a rDNA-derived organism or product.

Other Federal Agencies

Currently there is only one federal agency that has clear regulatory authority over a type of rDNA-derived product—the Environmental Protection Agency, under the Federal Insecticide, Fungicide, and Rodenticide Act (FIFRA). If someone developed a rDNA-derived microorganism for use as an insecticide, it would be regulated under this act. Further provisions have been proposed by EPA and other federal agencies to regulate rDNA products that fall under their authority.

Food and Drug Administration. The FDA would regulate rDNA products in the areas of drugs, medical devices, biologicals, animal drugs, food additives, and cosmetics. There have already been applications submitted that concern drugs and medical devices involving the use of insulin, interferon, and monoclonal antibodies. No food applicant and you will be doing yourself, your department, and your organization a big favor.

Training and Development. Without a successful training and development program, a company cannot meet its affirmative action goals. Every research manager, as well as every supervisor, needs to inform people of the training and education opportunities that relate to their work. A special effort should be made to encourage females and minorities to participate in the available programs.

The laboratory manager should delegate work and responsibilities to provide people with opportunities for growth and development. One way of achieving this is to rotate staff members' assignments to provide them with new experiences and the department with redundancy in skills. Obviously, this is more difficult in research, where many people have very specialized skills. However, many of the support assignments and administrative duties can be spread around. You need to set realistic but challenging objectives for all employees. Also, if you have trained your replacement there will be no delay in your own promotion should an opportunity present itself.

A good supervisor will provide frequent feedback about employee performance. A merit evaluation once a year is not often enough to help the employee achieve the development needed or deserved. Praise and recognition should be provided when earned, but do not hesitate to take corrective actions when necessary. Allow each employee enough freedom to make mistakes and to learn from their mistakes. Coach, counsel, challenge, lead, and motivate your employees. In other words, do whatever it takes to bring out the best in each of your people.

Career Planning. Not all employees are thinking about their career and its direction, but many are. Encourage them to discuss their career interests and aspirations and help them in any way you can to prepare for increased responsibility. Be realistic and honest with your people about career opportunities. Help arrange a transfer if a change is suitable to their skills, abilities, interests, and aspirations. Refer employees to a personnel specialist for assistance in career and educational counseling.

Discipline. People react in various ways to criticism and sometimes discipline involves termination. This is probably the most sensitive area in personnel relations. The laboratory manager must know the company rules and see that his employees know them also. An employee is entitled to know what is expected before being responsible for doing it. To accomplish this, all new employees should receive a copy of the company rules. When disciplinary action becomes necessary, follow the proper sequence: investigate the facts, give the employee your view of him, offer the employee a chance to respond, and then decide on and carry out the appropriate penalty.

Equitable treatment of all employees is essential; employee penalties for failure to comply should be weighted equally. Penalties which may be applied include, but are not limited to, the following:

- A verbal reprimand
- Written reprimand to the employee's personnel folder
- One or more days off without pay
- Demotion
- Termination if the employee is a habitual offender

Each manager should familiarize himself with his own organization's guidelines and policies with regard to penalties.

Be sure that you keep clear, accurate, and complete records. These are invaluable in establishing that fair and proper procedures were followed. In the absence of good records, most judgments have gone against the organization. Make every effort to avoid favoritism or even the appearance of it. A supervisor cannot be expected to like all employees equally, but he must treat them equally.

SUMMARY

The management of almost every type of scientific laboratory—whether academic, government, or proprietary, large or small—is likely to be affected at one time or another by government regulations. In recent years, the scope and extent of these regulations have evolved rather rapidly in response to a wide range of societal concerns centered on the safety of consumer products, the safety of the laboratory worker, and the protection of the environment. Regulations have also been developed to prevent questionable practices and the kind of fraud that has occasionally been committed in mandated laboratory studies in the past. Many of the newly promulgated regulations specify conditions under which laboratory studies must be conducted and thus affect the work itself, the process and/or procedures, and the final product. These regulations are overseen by a wide range of federal regulatory agencies, including the FDA, EPA, OSHA, NIOSH, USDA, DOT, NIH, and DEA. Because regulations continually undergo review and revision, the laboratory manager must remain aware and knowledgeable concerning the many government requirements that influence laboratory operation.

A good manager recognizes that highly trained and skilled employees are essential to the effectiveness of his operation. Although he is likely to be sensitive to individual aspirations and needs, personnel conflicts are inevitable. Just as regulations now govern the work process and product in the laboratory, regulatory statutes have evolved to protect the rights of the worker. Embodied in the Civil Rights Act of 1964 and the Affirmative Action Executive Order, these statutes establish guidelines by which an employer must deal with employees on a wide range of personnel practices. Because almost every personnel action has a potential legal result, it behooves the laboratory manager to be aware of the pertinent personnel regulatory guidelines and, above all, to be *fair* and *consistent* in the manner in which employees are treated.

REFERENCES

Broad, W., and N. Wade. 1982. *Betrayers of the Truth.* New York: Simon & Schuster.
Calvert, R. 1979. *Affirmative Action: A Comprehensive Recruitment Manual.* Garrett Park, Md.: Garrett Park Press.
Federal Register. 21 CFR. Updated regularly. Washington, D.C.: U.S. Government Printing Office.
Hayes, H. P. 1980. *Realism in EEO.* New York: John Wiley.
International Air Transport Association. 1985. *Dangerous Goods Regulations.* Montreal, Canada.
Krieg, N. R. 1984. *Bergey's Manual of Systematic Bacteriology.* Baltimore: Williams & Wilkins.
National Institutes of Health. 1979. *Laboratory Safety Monograph.* Rockville, Md.
_____. 1980. *Guidelines for Research Involving Recombinant DNA Molecules.* Rockville, Md.
National Institutes of Health and Centers for Disease Control. 1984. *Biosafety in Microbiological and Biomedical Laboratories.* Washington, D.C.
Occupational Safety and Health Administration. 1978. *General Industry Safety and Health Standards.* (29 CFR 1910); OSHA 2206. Washington, D.C.
U.S. Equal Employment Opportunity Commission. 1974. *Affirmative Action and Equal Employment—A Guide Book for Employers.* Washington, D.C.
U.S. Food and Drug Administration. 1982. *Toxicological Principles for the Safety Assessment of Direct Food Additives and Color Additives Used in Food.* Washington, D.C.
_____. 1983. *FDA Guidelines for Chemistry and Technological Data Requirements for Direct Food Additives and GRAS Food Ingredients.* Washington, D.C.

Appendix A
Research Involving Human Subjects

History is rich with examples of unethical human experimentation practices. One of the earliest recorded examples occurred in Persia, where a prince isolated newborn babies to determine whether language was a natural and spontaneous process. Later, Queen Caroline of England had the smallpox vaccination tested on orphans before administering it to her own children. And more recently participants in the Tuskegee study, who were mostly poor and black, were denied treatment for syphilis so that the natural course of the disease could be studied. There are many more documented cases of research involving the use of humans as experimental guinea pigs (Jonsen 1978; HEW 1973; Cardon, Dommel, and Trumble 1976).

It was the abuses of humans in research that occurred during World War II and the resulting Nuremberg war crimes trials that directed international attention to the ethical issues of human experimentation. Following these trials, the first written set of standards, "The Nuremberg Code of Ethics for Medical Research" was drafted (*U.S. v. Karl Brandt* 1949).

In the United States, twenty years elapsed before concern for the well-being of humans in research reached a point where something was done. Legislation was passed in 1974 establishing the National Commission for the Protection of Human Subjects of Biomedical and Behavioral Research under the auspices of the National Research Act.

The commission was charged with the responsibility to report on three general sets of topics and to make recommendations to both Congress and the Department of Health, Education, and Welfare

---

This appendix was written by Robert D. Colligan, president and chief executive officer of West Virginia University Medical Corp., Morgantown, WV.

(now the Department of Health and Human Services). These topics included (1) a proposal of guidelines for regulating research on specified populations; (2) a survey of the existing methods of ethical review of research involving human subjects in universities, hospitals, and other places receiving federal support for research projects, and a proposal of modifications deemed necessary; and (3) a report on the legal, social, ethical, and public policy implications of novel biomedical and behavioral technologies (Leske and Ripa 1980).

As a result of the commission's studies, recommendations for institutional review boards were published. Following the required period for comment on these recommendations, a list of regulations was proposed in August 1979. The regulations became effective in January 1981 (*Federal Register* 1981). They set the standard for moral and legal protection of human subjects involved in federally funded research.

INSTITUTIONAL REVIEW

Before 1960, about one-third of the medical schools had procedures to review research involving human subjects (Check 1980). The number of institutional review boards (IRBs) increased greatly when the Department of Health, Education, and Welfare required that institutions receiving federal health funds have review bodies. This requirement gained statutory power in the National Research Act of 1974.

The membership of the IRB is selected by the institution according to HHS guidelines. The IRB must have at least five members, one of whom is not affiliated with the institution. The membership may not be of one gender, race, or profession. Furthermore, each member must be sufficiently qualified to review research proposals emanating from the institution, but at least one member must be a nonscientist. The board may have representation from the clergy, law, or social work professions.

Institutional review boards are charged with the duty of assuring informed consent, protection of privacy and confidentiality, and adequate monitoring of research. Specifically, under the recommendation of the commission, IRBs shall determine that (1) the research methods are appropriate to the objectives of the research and the field of study; (2) the selection of subjects is equitable; (3) the risks to subjects are minimized by using the safest procedures consistent with sound research design and, whenever appropriate, by using procedures being performed for diagnostic treatment purposes; (4) the risks to subjects are reasonable in relation to anticipated benefits and the importance of the knowledge to be gained; (5) informed consent shall be sought under circumstances that provide sufficient opportunity for subjects to consider whether or not to participate and that minimize the possibility of coercion or undue influence; (6) informed consent shall be based upon communicating to subjects in language they understand information that they may reasonably be expected to desire in considering whether or not to participate; and (7) adequate provisions are taken to protect the privacy of subjects and to maintain the confidentiality of data.

INSTITUTIONAL RESPONSIBILITY

The institution has the responsibility to guide the IRB in its role of protecting human subjects and in so doing to recognize certain ethical responsibilities. In addition, the institution is charged with responsibility in the following areas (Wigody 1981):

1. Protection for subjects of non-federally funded research. The new regulations state that the institution's statement of intent to the secretary of HHS shall include a statement of principles "governing the institution in the discharge of its responsibilities for protecting the rights and welfare of human subjects of research conducted at or sponsored by the institution regardless of the funding source."
2. Research exempt from IRB review. Certain categories of research are exempt from IRB review, mostly in the areas of behavioral research. Other areas include the collection or study of existing data, documents, records, pathological specimens, or diagnostic specimens, if these sources are publicly available or if the information is recorded by the investigator in such a manner that a subject cannot be identified, directly or through identifiers linked to the subjects. Knowing the definitions, it is up to the researcher and the institution to decide if the project is exempt, and whether it should be reviewed by the IRB.
3. Monitoring. The institution shall assure that the IRB reviews and approves proposed changes in ongoing research protocols before initiation of these changes occurs.
4. Projected design and methodology. The IRB is not required to determine that research methods are appropriate to the objectives of the research and the field of study. Rather, the IRB shall determine that the risks to subjects are minimized. However, the institution may have responsibility to protect subjects from participating in research of poor design.
5. Civil liability of physician and institution. In its grant of author-

ity to the Food and Drug Administration (FDA) to promulgate regulations to protect public health, Congress has not specifically provided the FDA with the power to disqualify an investigator or to suspend a study before or after disqualification. However, authority not provided by the law is often assumed by regulatory agencies within the government (Lasagna 1980).

DEFINITIONS

Human Subjects

Any individual who participates in a research project and from whom data is obtained through intervention or interaction with an investigator is a human subject. This includes inpatients and outpatients; donors of organs, tissues, and services; informants; normal volunteers including students, prisoners, residents, and clients of institutions for the mentally ill and mentally retarded; and persons in the military. Also included are the newborn and the dead, whose rights may be expressed by the next of kin.

Children

Special regulations exist for the use of children in research. In order for the the Department of Health and Human Services to approve funding, the IRB must find that (1) any risk represents a minor increase over minimal risk, (2) the intervention is likely to yield knowledge that is of vital importance for understanding or relieving the subject's disorder or condition, and (3) proper provisions have been made for attaining the consent of the child and parent or guardian.

The HHS will fund research that does not meet the above requirements only if (1) the IRB finds that the research presents a reasonable opportunity to further the understanding, prevention, or alleviation of a serious problem affecting the health or welfare of children; and (2) permission is obtained from the secretary of HHS after consultation with a panel of experts.

REVIEW PROCESS

The review of a research proposal begins with the determination of whether human subjects are exposed to risk. If it is found that the proposed procedures are established, accepted, and necessary to the needs of the subject, the proposal is approved, and the review ends.

If the review determines that the procedures to be applied will place the subject at risk, then the relative risk-benefits ratio is considered, as well as the subject's rights. The board reviews the provisions for adequate protection of the subjects from risks and ensures that there are provisions for adequate and appropriate consent procedures. The board also ensures that information that could be traced to or identified with the subjects is kept confidential.

Research that does not initially involve human subjects but may involve them after the successful completion of preliminary studies should be reviewed and certified in the above manner. These projects are reviewed again prior to the beginning of the budget period during which human experimentation is expected to begin.

Expedited Review

An expedited review may be made for categories of research that do not require a consent form. The chairman of the IRB or his designate committee may review and approve the following:

A. Survey or interview procedures except where
 1. subjects can be identified
 2. responses could place the subject at risk
 3. research deals with sensitive aspects of behavior such as illegal conduct or sexual behavior

B. Observation of public behavior except where
 1. subjects can be identified
 2. subjects could be placed at risk of criminal or civil liability
 3. the research deals with sensitive topics

C. Study of data, documents, records, and specimens

D. Research involving only minimal risk

E. Minor changes in proposals that have been previously approved by the IRB

Continuing Review

During the initial review, time intervals for continuing review are determined by the board. The frequency of intervals is related to the project's risk. Each project must be reviewed at least annually to permit certification of review on noncompeting continuation applications.

Decision

The IRB notifies the investigator and the institution, in writing, of the decision to approve, disapprove, or modify a research protocol. If the IRB disapproves the protocol, it must include a statement of reasons. The investigator then has the opportunity to respond. Appeal of unfavorable recommendations, restrictions, or conditions can be made to the board or to another appropriate review group.

Documentation

The board is further responsible for maintaining copies of all documents presented for the initial and continuing reviews, including transmittal of action, instruction, and conditions resulting from deliberations during review. These records must be retained for three years after completion of the research.

Enforcement

The staff of HHS may investigate the board and its records from time to time. If there is evidence that the institution has failed in its duties to ensure the protection of rights and welfare of research subjects, whether or not HHS funds have been used, the staff can recommend that the institution and the individuals concerned no longer receive future funds. This means that the board must report noncompliance of investigators through the appropriate channels.

INFORMED CONSENT

The definition of informed consent is "the knowing consent of an individual or his legally authorized representative, so situated as to be able to exercise free power of choice without undue inducement or any element of force, fraud, deceit, duress, or other form of constraint or coercion" (Leske and Ripa 1980). The United States regulations for assuring informed consent derive from the Nuremberg Code of Ethics for Medical Research. The new federal regulations have expanded the scope and are specific as to its elements.

Type I

The consent form must contain the following:

1. A general description of the study and its purpose, followed by a fair and clear explanation of all procedures that will be used for experimentation and treatment, including those that are experimental
2. An explanation of the conditions for participation and an assurance that the identity or particular finds on each subject will be held confidential by the investigators
3. A description of any known attendant discomforts or risks that are reasonable to expect
4. A description of the benefits that are reasonable to expect
5. A disclosure of appropriate alternative procedures that might be advantageous for the subject, rather than the procedures under investigation
6. An offer to answer any inquiries concerning the study prior to its inception or during its course
7. An explanation of the availability of treatment if injury occurs and the availability of compensation if injury occurs
8. An instruction that the subject is free to decline participation, withdraw consent, or discontinue participation at any time and without prejudice

In addition, subjects must be given the name of a person to contact for answers to questions or in the event of a research-related injury or illness.

The subject or his authorized representative signs this form and retains a copy. This consent form is required in all cases of significant or unknown risks and in proposals using volunteers, prisoners, and other individuals having limited freedom and mental abilities.

Type II

A second type of consent form can be used in proposals with average or usual risks. It consists of a short written document indicating that the basic elements of informed consent presented above have been delivered orally to the subject. The short form must also be signed by the subject or his representative, and by a witness to the oral presentation. Sample copies of the written consent form and summaries of oral statements that have been approved by the board are to be retained in the board's files. A copy is also given to the subject.

Type III

A third type of consent is a modification of the above forms that may be used when the investigator can prove that risk to any subject is minimal and that use of either of the other forms would invalidate objectives of immediate importance or be less advantageous to the subject.

Children

Parents or guardians must consent to the use of their children in research except where (1) research is conducted in commonly accepted educational settings and involving normal educational practices, or (2) research involves the observation of existing data, documents, records, or specimens.

SUMMARY

The National Commission for the Protection of Human Subjects of Biomedical and Behavioral Research recognizes that research with human beings is essential for fighting disease and in expanding the frontiers of knowledge. It is not the government's intention to police or regulate the type of research that is conducted in this country, but rather to assure that this activity is carried out without needless risk of distress and with the willing and knowing cooperation of its subjects (Jonsen 1978).

The commission places the onus of this responsibility with the institution sponsoring the research. Through its institutional review board (IRB), the institution interpets the regulations, reviews research protocols involving human subjects, and makes judgments on whether the amount of risk to the subject is justifiable by the impact of the study on the body of scientific knowledge.

The Department of Health and Human Services (HHS) has been granted authority to review ongoing research that has been previously approved, to ensure that standards relating to human experimentation are maintained. Occasionally violations have been found, but the method for reporting accidents or investigator noncompliance is inconsistent and the sanctions imposed on noncomplying investigators and their sponsoring institutions also lacks uniformity.

Congress has mandated that the commission report every two years on the degree to which federal departments are implementing the regulations for the protection of human subjects in research. Specifically, the commission will make regulations that attempt to provide (1) uniformity in the regulations for the protection of human research subjects among the several federal agencies, (2) standards for handling reports of harm or misconduct involving human subjects, and (3) procedures for collecting and reporting the number of subjects adversely affected by each research protocol.

REFERENCES

Cardon, P. V., F. W. Dommel, and R. R. Trumble. 1976. "Injuries to Research Subjects." *New England Journal of Medicine* 295:650-54.
Check, W. A. 1980. "Protecting and Informing Human Research Subjects." *Journal of the American Medical Association* 234:1985-93.
Control Council Law No. 10. International Military Tribunal. *Trials of War Criminals before the Nuremberg Military Tribunals:* United States v. Karl Brandt. 1949. 2:181-83. Washington, D.C.: U.S. Government Printing Office.
Federal Register. 1981. Vol. 46. Washington, D.C.: U.S. Government Printing Office.
Jonsen, A. R. 1978. "Research Involving Children: Recommendations of the National Commission for the Protection of Human Subjects of Biomedical and Behavioral Research." *Pediatrics* 6:131-36.
Lasagna, L. 1980. "Regulation and Constraint of Medical Research: Rights and Human Welfare." *American Review of Respiratory Diseases* 122:361-64.
Leske, G. S., and L. W. Ripa. 1980. "Ethical and Legal Considerations Associated with Clinical Field Trials." *Journal of Dental Research* 59:1243-53.
U.S. Department of Health, Education and Welfare, Public Health Service. 1973. *Final Report of the Tuskegee Syphilis Study Ad Hoc Advisory Panel.* Washington, D.C.: U.S. Government Printing Office.
Wigodsky, H. S. 1981. "New Regulations, New Responsibilities for Institutions." *Hastings Center Report* 11:12-14.

Appendix B
Research Involving Animals

The use of animals as research subjects has a history that dates back to mankind's first inquiries into the mysteries of anatomy and physiology. On the other hand, those who have questioned the use of animals in demonstrations and experimentation have a history just as long. Regardless of the antiquity of the opposing views, the underlying tenets have changed little (Sperlinger 1981, Miller and Williams 1983). The major arguments against the use of animals in research are fourfold:

1. Regardless of how important the goal of the experiment may be, the infliction of pain upon animals is unjustified.
2. Only small benefit has been derived from animal experimentation—most experiments have little relevance to humans.
3. Most animal experiments now conducted could be eliminated by using alternative methods.
4. Humans have no right to use animals for the sake of achieving human ends.

The proponents of animal research retort:

1. Animal experimentation is aimed at increasing human knowledge and/or welfare.
2. Most experiments involve no serious pain and/or discomfort to animals and scientists strive to minimize pain.
3. At present few alternative techniques are available that can replace animal experimentation and alternative techniques are used when possible.
4. Man has the right to use animals in pursuit of important human goals.

Most vivisectionists justify their position by citing the many medical advances that have reduced human mortality and morbidity rates (Paton 1979).

Proponents of both views seem to be even more polarized today than ever—with a great deal of emotionalism welling forth from the extreme fringes of both groups. Nevertheless, humane care and use of laboratory animals has become an issue of broad societal concern and undoubtedly the more moderate voices of the animal rights movement have increased researchers' awareness of the need for humane treatment of animals and have contributed to improvement in laboratory animal care.

THE HISTORY OF ANIMAL WELFARE LEGISLATION

The earliest effective reform movement concerning the use of animals in research and demonstrations occurred in the mid-nineteenth century in England. This led to the passage, in 1822, of the first comprehensive legislation to protect animals used for research purposes. This early legislation was replaced in 1876 with the Cruelty to Animals Act, which has remained in effect to the present. The only experiments permitted under the act, and then only under license from the Home Secretary, are those performed with the intent of advancing physiological knowledge or knowledge thought useful for saving or prolonging life or alleviating suffering in man or animals. No experiments are permitted for the sole purpose of obtaining manual skill. Under license, every experiment must be performed using an anesthetic throughout, and the animal must be killed at the end of the experiment while under the anesthetic. Experiments must not be performed as a demonstration in lectures in medical schools, hospitals, colleges or elsewhere. A licensee may be released from the specific restrictions listed above, dependent upon the skill and/or training of the individual, the experimental procedure, and species of animal used, by obtaining an appropriate certification. Experimental procedures considered in the licensing process and requiring certification include four categories: experiments conducted (1) under anesthesia without recovery, (2) under anesthesia with recovery, (3) with no anesthesia, and (4) lectures and demonstrations under anesthesia without recovery.

The Cruelty to Animals Act is not only a landmark in animal welfare legislation but has served as framework to subsequent regulatory legislation in other countries. Currently, a Committee of Experts on

Animal Protection, under the aegis of the Council of Europe, is attempting to frame a European Convention on the Protection of Laboratory Animals. If this effort is successful, it is likely that many of the principles expressed in existing British legislation will be adopted.

It was nearly 100 years after the British Cruelty to Animals Act was implemented before animal welfare legislation was passed in the United States. The Laboratory Animal Welfare Act (9 CFR Subchapter A) was enacted by the U.S. Congress in 1966 and amended in 1970 and 1976. Basically the act invests the U.S. Department of Agriculture's Animal and Plant Health Inspection Service (USDA/APHIS) with responsibility for the issuance and enforcement of regulations regarding humane care, handling, treatment, transportation, general husbandry (housing, sanitation, and feeding), employee training, veterinary care, quarantine, and separation of species. The animal welfare act also requires registration of research facilities and licensing of laboratory animal dealers.

Although the act does not affect the design, procedures, or performance of the actual research, it does require every research facility to show "that professionally acceptable standards governing the care, treatment and use of anesthetic, analgesic and tranquilizing drugs, during experimentation are being followed by the research facility during actual research or experimentation."

As part of the enforcement of the animal welfare act, the USDA requires annual reporting from each registered research facility using live animals in research or testing. Although the act does not require inspection of government research facilities, federal agencies are directed to comply with its provisions and hence must also submit annual reports. The report must include the numbers of animals (by species) used in experimentation. These numbers must be further separated into three categories of experimental use: (1) no pain or distress, (2) pain or distress alleviated by the use of appropriate drugs, and (3) pain or distress without the use of appropriate drugs.

In addition to the Laboratory Animal Welfare Act, two other laws have been enacted to protect animals used in research. These are the Endangered Species Act (50 CFR 17) and the Marine Mammal Protection Act (9 CFR subchapter A). Both laws govern the purchase, acquisition, or capture of species falling within their respective purview. Special permits for use of such animals must be obtained. Scientists and public display facilities are the only eligible recipients of such permits. The Endangered Species Act further restricts the scope of permissible experiments to studies that would directly benefit the species under investigation.

INDUSTRIAL RESEARCH

Of the approximately 70 million animals currently employed in research in the United States, about twenty percent are required for toxicity testing. Largely resulting from the legal obligations imposed on industry by enactment of federal legislation, these procedures involve the testing and evaluation of substances that could be potentially harmful to humans, animals, and the environment. The principal regulations are summarized in table B-1. A more complete compilation of federal statutes governing the regulation and control of toxic substances may be found in *An Analysis of Past Federal Efforts to Control Toxic Substances,* published by the Environmental Law Institute, Washington, D.C. (Doniger, Liroff, and Dean 1978). Whereas these regulations cover a wide range of substances and products that must be tested for toxicity and are embodied in a wide-ranging number of statutes, the enactment of the Toxic Substances Control Act (TSCA) in 1976 has brought a large number of compounds that require testing under one legislative umbrella. TSCA, for which the EPA is responsible, requires testing and submission of test results before the production of the respective chemical substances can begin. It includes *all new* chemicals with the exception of food additives, drugs, pesticides, alcohol, and tobacco.

It is imperative that managers of animal research laboratories have a thorough knowledge of the federal regulations affecting their specific operation. In addition, it should be pointed out that the *good laboratory practice* regulations for nonclinical laboratory studies (21 CFR 58), administered by the FDA, also impose stringent require-

Table B-1. Principal Federal Statutes Affecting Testing and Control of Toxic Substances

Statute	Responsible Agency
Food, Drug, and Cosmetic Act, 1938	FDA
Public Health Service Act, 1944	FDA
Federal Insecticide, Fungicide, and Rodenticide Act, 1954, amended 1972	EPA
Federal Meat Inspection Act, 1967	USDA
Occupational Safety and Health Act, 1970	NIOSH/OSHA
Federal Water Pollution Control Act, 1972, amended 1977	EPA
Safe Drinking Water Act 1974, amended 1977	EPA

ments with regard to animal care, animal care facilities, animal supply facilities, laboratory operation areas, protocol documentation, records, and reporting.

Notice of the number of legislative acts impacting animal research makes one point obvious: those whose research requires animals are caught in the middle of two strong opposing societal forces. On the one hand are the animal rights groups who exert pressure for the reduction or abolition of animal experimentation, and on the other is the demand for safe, effective consumer products and an unspoiled environment. Products such as soaps, toothpastes, shampoos, mouthwashes, and baby products are applied to the skin, eyes, and other sensitive areas of the body. The purity of individual constituents and safety of the product must be verified—requiring, in almost every instance, testing in animals. Logic dictates that most animal research must continue. However, laboratory managers should be aware of certain types of experiments, notably toxicity tests, that deserve closer scrutiny, particularly the LD_{50} and Draize irritancy tests.

The LD_{50} represents the dose of a substance required to kill 50 percent of the test subjects within fourteen days. This test was originally devised to provide an accurate method for measuring potency of highly toxic but efficacious medicinal drugs. In the case of agents such as digitalis and insulin, for example, where the difference between beneficial and dangerous doses is small, the strength of the drug must be known precisely. The degree of statistical precision required, usually within the 95 percent confidence limits, necessitates the testing of large numbers of animals. In many instances, however, more precise biochemical assays have been developed that eliminate the necessity for any other than semiquantitative parameters—that is, the general order of magnitude of acute toxic doses. This information might now be derived from a smaller number of animals such as that used in "limit tests" or tests of the approximate lethal dose (ALD). Recently, the FDA, with major responsibility for safety testing of drugs, has altered its policy requiring LD_{50} tests, although references to required LD_{50} are still found in older guidelines. The requirement remains for certain highly toxic antitumor antibiotics with narrow margins of safety. It seems certain that other federal agencies still requiring the LD_{50} will revise their policies in the near future.

The Draize opthalmic irritancy test estimates irritation by a test substance based upon the degree of corneal opacity and conjunctival inflammation in test animals—usually rabbits. In many cases these tests cause severe pain to the test animal and interpretation of test results is somewhat subjective. Although the Draize test is proving difficult to replace with alternative methods, as employed today many of the stressful features of the original test are avoided by using more dilute forms of the test substance. Nevertheless, this test represents another instance where prudence should be exercised in decisions regarding the necessity for obtaining this type of toxicity data.

BIOMEDICAL RESEARCH

The National Institutes of Health (NIH), including its extramural grant and contract programs, account for by far the largest number of animals used in research in the United States. The demand for research animals reflects increased public concern over health matters, especially cancer and other major diseases. At the same time this explosion in legislatively directed monies into health research programs occurred, the NIH became acutely aware of the need for policies governing the care and use of laboratory animals in biomedical research. NIH first published the *Guide for the Care and Use of Laboratory Animals* in 1963, three years prior to enactment of the Laboratory Animal Welfare Act (NIH 1985). The purpose of the guide is "to assist scientific institutions in providing professionally appropriate care for laboratory animals" and to serve as the standard for care and use of animals at facilities where NIH-sponsored programs are located.

In addition to adherence of the animal care and use principles provided in the guide, institutions receiving NIH support are required to file an "Animal Welfare Assurance Statement" with the Office for Protection from Research Risks (OPRR), NIH, as directed by Public Health Service (PHS) animal welfare policy. Further, NIH requires that its grant or contract awardee institutions have animal care committees whose purpose is to provide consultation to both institutional administration and individual investigators with regard to the care and use of laboratory animals.

Due to concern about the adequacy of the assurance program, NIH undertook site visits to selected recipient institutions in 1983. These inspections resulted in recommended changes in PHS Animal Welfare Policy—specifically in regard to the composition and function of the institutional animal research committee (NIH 1984). In addition, NIH has recognized accreditation by the American Association for Accreditation of Laboratory Animal Care (AAALAC) as a means of achieving and maintaining animal care standards that are commensurate with an acceptable institutional assurance.

AAALAC was founded to conduct a voluntary program for the accreditation of laboratory animal facilities and to encourage, promote, and facilitate scientific research. AAALAC encourages high standards (the program uses the NIH guide as its primary reference standard) for the care and use of experimental animals, including appropriate veterinary care, controlling variables that might adversely affect animal research, and protecting the health of animal research personnel. A council on accreditation conducts site visits to institutions, evaluates the site visit reports, and makes recommendations concerning the accreditation status of the facility and its programs. NIH encourages AAALAC accreditation by accepting it as one of two options in achieving an assurance acceptable to OPRR.

SUMMARY

The broad spectrum of legislative edicts that govern the use of laboratory animals for research are continually undergoing review and revision. Many federal agencies now cooperate through a newly established Interagency Research Animal Committee (IRAC) and through interagency agreements in order to focus on the issues involving animals used in testing and biomedical experimentation.

All persons involved in animal research must take the initiative in keeping themselves informed of the latest concerns and policy affecting the conservation, use, care, and welfare of research animals. The judicious use of animals in research, and particularly concern for their care and treatment, will ultimately be a reflection upon your ability to manage a most valuable research resource.

REFERENCES

Doniger, D. D., R. A. Liroff, and N. L. Dean. 1978. *An Analysis of Past Federal Efforts to Control Toxic Substances.* Washington, D.C.: Environmental Law Institute.

Miller, H. B., and W. H. Williams, eds. 1983. *Ethics and Animals.* Clifton, N.J.: Humana Press.

National Institutes of Health. 1984. *Laboratory Animal Welfare: NIH Guide for Grants and Contracts,* NIH 13:1-27. Bethesda, Md.

———. 1985. *Guide for the Care and Use of Laboratory Animals.* DHHS (NIH) Publication no. 85-23. Bethesda, Md.

Paton, W. D. M. 1979. "Animal Experiment and Medical Research: A Study in Evolution." *Conquest* 169:1-14.

Sperlinger, D., ed. 1981. *Animals in Research: New Perspectives in Animal Experimentation.* New York: John Wiley.

Index

Abeles, F. B., 146, 153, 155
Accident investigation, 67-69. *See also* Laboratory safety
Actual reduction to practice, 116-17
Acute hazardous wastes, 191
Adjacent homolog, 109-10
Administrative reports, 89
Aerosoling, 79
Affidavits for patent procurement, 110, 114
Affirmative action. *See* Equal employment opportunity and affirmative action
Age Discrimination Act of 1967, 210
Agriculture Department, U.S., 195
Allowable costs, 164-65
American Association for Accreditation of Laboratory Animal Care, 232-33
An Analysis of Past Federal Efforts to Control Toxic Substances, 230
Animal and Plant Health Inspection Service, 229
Animal research subjects
 for biomedical research, 232-33
 FDA requirements, 184
 grant proposals involving, 144, 148-49
 for industrial research, 230-32
 legislation regarding, 228-29
 overview of, 227-28
Animal Studies Committee, 144
Animal Welfare Assurance Statement, 149, 233
Applied science, 105
Appropriations Committees, House and Senate, 131
Atomic bomb, 126
Audiovisual materials for technical communication, 92-94

Badawy, M. K., 8
Barkley, W. E., 75, 76, 83
Bid solicitation procedures, 175-76
Biles, B. R., 141
Biocontainment stations, 75, 76
Biohazard hoods, 74-76
Biohazards, 81-82
Biological safety cabinets, 75, 76, 82
Biomedical research support grants, 139, 140
Biomedical research using animal subjects, 232-33
Biosafety levels, 82, 83
Black, H. S., 162
Boots, protective, 72, 74
Booz, Allen and Hamilton, 47
Bottle carriers, 74
Bowden, J., 35
Break-even point (BEP), 52-53, 57
Broad, W., 183
Brooks, H., 126
Budget Committees, House and Senate, 130-31
Budget preparation as part of proposal preparation, 150-53
Bush, Vannevar, 126

Canada, patent applications, in, 115
Cangemi, R. R., 5
Capital equipment, 25
Cardon, P. V., 217
Career planning, 214
Carlson, Chester, 105
Cartoons, 94
Catalog of Federal Domestic Assistance, 138, 142
Center grants, 140-41
Chargaff, E., 127

235

Charts, 93
Check, W. A., 218
Chemical industry, R&D funding in, 10
Chemicals. *See also* Hazardous materials; Laboratory safety
  corrosive, 78-79
  disposal of, 81
  inventory system for, 78
  patent issues regarding, 109-11
  storage of, 77-78
  Transportation Department guidelines on, 79-80
Children used in research studies, 220
Civil Rights Act of 1964, 210
Code of Federal Regulations, 80
Cole, S., 157
Colligan, Robert D., 217*n*
Commerce Business Daily, 176
Committee of Experts on Animal Protection, 229
Committee on Science and Astronautics, House, 131
Communication, 95-96. *See also* Technical communication
Competitive bids, 175-76
Components, 203
Computerization of project management techniques, 15
Confidential information, 102-4
Congressional Budget Act of 1974, 128, 129
Congressional Budget Office, 131
Congressional review as step in federal budget cycle, 130-31
Consortium grants, 139
Constitution, U.S., as basis for patent and copyright law, 104-5
Constructive reduction to practice, 116-17
Contact lenses, 74-75
Containment. *See also* Hazardous materials
  of biohazardous material, 81-83
  of recombinant DNA, 198-99
Contracts
  bid solicitation procedures, 175-76
  decision criteria, 176-77
  grants and, 138, 173-74
  information on availability of, 176
  management of, 178
  proposal preparation, 177-78
  responsibilities in conducting research under, 181
  sponsoring, 178-81
  types of, 174
Contractual services, outside, 24-25

Controllable costs, 22
Controlled substances regulation, 199-201
Control process
  elements of, 42-44, 56-57
  importance of, 41-42
Control techniques
  financial, 50-53
  specialized objective, 44-50
  subjective, 53-54
Copyright law, 104-5
Corrosive chemicals, 78-79. *See also* Hazardous materials
Corrosivity, 189
Cost-plus contract, 174
CPM (Critical path method), 47
Critical path, 13, 14, 47
Cruelty to Animals Act, 228
Curriculum vitae, 24

Dangerous Goods Regulations, 193
Davies, R. K., 62
Day, R. A., 90
Dean, B. V., 50, 54, 56
Dean, N. L., 230
Decision making, 35-38
Decision tree, 37, 38
Delphi forecasting technique, 55
DELTA charts, 14, 16
Departmental overhead, 22
Diamond v. Charkrabarty, 106-7
Direct costs
  expenditures report, part of, 22
  grant proposals, part of, 150-53
Direct project support, 25
Discipline, employee, 214-15
Discretionary grants, 139
Discrimination. *See* Equal employment opportunity and affirmative action
Division of Research Grants, 157
Dommel, F. W., 217
Doniger, D. D., 230
Draize tests, 231-32
Drug and device approval
  general provisions regarding, 206-7
  overview of, 205-6
  and quality assurances, 207-8
Drug Enforcement Agency (DEA), 199-201
Duga, J. J., 127
Dusting, 79

Eaves, G. N., 146, 157-59, 161
Edelman, B., 176
Edelman, J., 177

Education/training grants, 140
Emergency equipment, 72-73
Employees. *See* Personnel
Employment interviews, 213-14
Endangered Species Act, 229
Environmental Protection Agency
  (EPA), 179
  and hazardous wastes, 188
  laboratory practice regulations, 185-86
  recombinant DNA regulated by, 194-95
Epperly, W. R., 27
EP toxicity, 190
Equal employment opportunity and
  affirmative action
  company's role in, 212
  history of, 209-10
  law into practice, 210-11
  present status, 211-12
  supervisor's role in, 212-15
Equal Employment Opportunity
  Commission (EEOC), 210
Equipment
  budgeting and grant proposals for, 152
  emergency, 73-74
  protective, 74-75
European Convention on the Protection
  of Laboratory Animals, 229
European Patent Convention, 108
Evaluation methods, 43
Expenditures
  changes in, 164-65
  reports of, 165-66
Expense spending reports, 22, 23
Exploratory forecasting, 55
Eyewear, 72, 74

Facilities assistance grants, 140
Facility resources, 26
Fair Play Act of 1963, 210
Farbenfabriken of Elberfeld Company
  case, 106
Fawcett, H. H., 69
Fayol, Henri, 41, 43
Federal budget cycle
  budget execution, 131-32
  budget formulation, 128-30
  congressional review, 130-31
  impediments to, 134-35
  table of steps, 133
Federal Contracts, 176
Federal Food, Drug, and Cosmetic Act,
  184, 199, 205
Federal funding for research, 125-28,
  137, 138. *See also* Federal budget
  cycle; Grant proposal process;
  Grant management; Grants

Federal Hazardous Substances Act
  (FHSA), 179
Federal Insecticide, Fungicide, and
  Rodenticide Act (FIFRA), 178,
  179, 185, 194
Federal Register, 138, 142, 185, 186, 194
Feedback systems, 43
Financial control techniques, 50-53
Financial resources, 21-23. *See also*
  Research funds
Financial source summary, 162-64
Financial statements, 51-52
Financial status reports, 166
Fire emergency training program, 74
Fire extinguishers, 74
Fiscal policy, 134-35
Fisher, W., 127
Fixed costs, 52-53
Fixed fees, 174
Flammable materials
  disposal of, 81
  storage of, 77-78
Flip charts, 92
Followers, 32-33
Food additives (GRAS) regulation, 201
  on components, 203
  and final product evaluation, 203-5
  on microorganisms, 202-3
  and petition preparation, 205
Food and Drug Administration (FDA),
  184, 185
  on drug and device approval, 205-8
  on food additive approval, 201-5
  on human research subjects, 220
  on LD 50 tests, 231
  on recombinant DNA, 194-95
Formula grants, 139
Forward funding, 134
Foster, R. N., 36, 37
Foundation Center, 142
The Foundation Directory, 142
The Foundation Grants Index, 142
Francis, P. H., 49
Fringe benefits, 151-52
Fume hoods, 74, 75
Furnas, C. C., 89, 96
Fusfeld, A. R., 55

Gantt, Henry, 44
Gantt techniques, 13-15, 44, 45
Gas masks, 74
General Industrial Safety and Health
  Standards, 186
GERT (Graphical Evaluation and
  Review Techniques), 47
Giegold, W. C., 35

Glove boxes, 74
Goggles, 72, 74
Good laboratory practice (GLP) guidelines, 181
Good Laboratory Practice (GLP) Regulations for Nonclinical Laboratory Studies, 184, 230-31
Good laboratory practices (GLPs), 183-89. *See also* Laboratory management practice regulations
Governmental regulation compliance, 183
 laboratory management practices in various agencies and, 183-88
Gowing, M., 126
Grant application kits, 143
Grant management
 changes in expenditures and costs, prior approval for, 164-65
 expenditure reports, 165-66
 invention reports, 167
 overview of, 161-64
 progress reports, 166-67
 transfers, termination, and special circumstances, 167-68
Grant proposal process, 11-12
 budget preparation, 150-53
 first steps, 141-45
 preproposals, 143-44
 proposal preparation, 145, 146
 research plan, 146-50
 review and award system, 156-61
Grants
 auditing of, 168
 contracts and, 138, 173-74
 definition of, 137
 differences in, 137-39
 disapproval of, reasons for, 154-56
 federal, 137, 138
 review and award process, 156-61
 transfer or termination of, 167
 types of, 139-41
Grantsmanship Center training programs, 144
Grant sources, 141-42
Grants Weekly, 176
Graphics, 92-94
Graphs, 94
GRAS. *See* Food additives (GRAS) regulation
Guide for the Care and Use of Laboratory Animals, 149, 232
Guidelines for Research Involving Recombinant DNA Molecules, 194
Gulley, G., 177

Hazardous materials, 75-77
 handling of, 78-79
 pollution regulation and, 192
 safety equipment for working with, 74-75
 storage of, 77-78
 transportation of, 79-80, 189, 192-93
Hazardous wastes
 definition and criteria for, 189-90
 disposal of, 80-81, 190-92
 governmental regulations regarding, 188-93
 transportation of, 189
Health, Education, and Welfare Department, U.S., 217-18
Health and Human Services Department, U.S., 218-20, 224
Hensley, O., 177
Herzberg, F., 29
Herzberg's motivational theory, 29, 31
Hierarchy of needs, 29, 30
Hodgetts, R. M., 41, 43
Hoods, 74-76
Human research subjects
 background of, 217-18
 definitions of, 220
 grant proposals for projects involving, 144, 148-49
 informed consent requirement, 222-24
 institutional responsibility for, 219-20
 institutional review regarding, 218-19
 review process regarding, 220-22
Human resources. *See* Personnel
Human Studies Committee, 144
Humorous sketches, 94

Ignitability, 189
IMPACT (Implementation Planning and Control Technique), 47
Independent events, 13
Indirect costs
 in expenditure reports, 22
 in grant proposals, 153-54
Indirect project support, 25-26
Individual researchers, 27-28
Industrial research
 animal subjects used in, 230-32
 funding in, 10
Informed consent, 149
 of children, 224
 definition of, 222
 types I-III, 223-24
Institutional biosafety committee (IBC), 196, 197
Institutional review boards, 218-19

Insurance benefits, 151-52
Intellectual property. *See* Proprietary information
Interference procedure regarding patent application, 109, 115-20
Intermediate-range planning, 10-12
International Directory of Contract Laboratories, 180
Interviewing of job applicants, 213-14
Invention reports, 167
Inventorship, 112
Inventorship awards, 121
Invitation to bid (ITB), 175, 176
Irving, J. B., 62, 64

Jackson, B., 36, 37
Jackson, E. M., 180
Jagger, J., 146, 149
Jeffers, Jerome L., 100*n*
Joint Economic Committee, 130
Jonsen, A. R., 217, 224
Journals, 88-90

Kapista, P., 127
Kenny, J. T., 162
Kepner, C., 34
Kewanee Oil Company, 102-4
Key result areas (KRAs), 9-10
Kiritz, N. J., 144, 146, 149
Krieg, N. R., 202

Laboratory Animal Welfare Act, 229, 232
Laboratory design, 70-73, 82
Laboratory management practice regulations
  on controlled substances, 199-201
  on dress and device approval, 205-8
  of EPA, 185-86
  of FDA, 184, 201-8
  on hazardous substances, 188-93
  of NIOSH, 187-88
  of OSHA, 186-87
  on personnel practices, 209-15
  on recombinant DNA, 194-99
Laboratory notebooks, 120
Laboratory safety
  biohazards and, 81-83
  committee for, 66-67
  emergency equipment, 73-74
  hazardous materials use and, 75-81
  laboratory design and, 70-73
  management of, 61-70
  program for, 64-66
  radiation hazards and, 82-83
  special equipment and facilities for, 74-75
  working alone and, 83-84
Laboratory Safety Monograph, 197
LD 50 tests, 231
Leavitt, H. J., 33
Legislative Committees, House and Senate, 131
Leske, G. S., 218, 222
Letters of intent, 143
Lewis, D. V., 91, 95
Licensing technology, 120-21
Lin, W. T., 50, 51
Lirofff, R. A., 230
Loners, 32
Long-range planning, 4-5
Loperfido, J. C., 66

McCloughlin, W. G., 49
McConkey, D. D., 8, 9
McGregor, Douglas, 29, 31
McGregor's Theory X, Theory Y, 29, 31, 32
Management by Objectives (MBO), 8-9
Management by Objectives for Results (MBOR), 9
Management by Results (MBR), 8, 9
Manhattan Project, 126
Manpower. *See* Personnel
Marine Mammal Protection Act, 229
Maslow, A. H., 29, 30
Maslow's hierarchy of needs, 29, 30
Matrix analysis, 5, 6
Matrix management, 27, 28
Mausner, B., 29
Media, technical communication and, 97
Medical Device Amendment of 1976, 184
Merrifield, D. B., 36-38
Merritt, D. H., 159
Microorganisms, 202-3
Milestone analysis, 44-46
Miller, H., 35
Miller, H. B., 227
Morrisey, G. L., 9
Motivation
  individual types and, 32-33
  position design as element of, 31-32
  recognition as form of, 33
  theories of, 29-31
Murry, J. P., 141
Murtaugh, J. S., 125

National Advisory Council, 157-59
National Commission for the Protection of Human Subjects of Biomedical and Behavioral Research, 217, 224
National Fire Protection Association (NFPA), 72
National Institute for Occupational Safety and Health (NIOSH), 187-88, 195
National Institutes of Health (NIH), 125
and animal research subjects, 232
grant and contract guide, 138
grant application format for, 146-52
grant management requirements of, 164-68
grant proposal rejection, study on, 154-56
grants provided by, 141
Office for Protection from Research Risks (OPRR), 149, 232, 233
peer review for grant proposals, 157-60
recombinant DNA guidelines, 194-99
study section reviews, guide for, 144
National Research Act, 217, 218
National science policy, 125-28
Network analysis, 12-13, 47-50, 56
Network switching, 50
Nixon, R. A., 22
Normative forecasting, 55
Notice of Grant Award, 159
Nuremburg Code of Ethics for Medical Research, 217, 222

Objectives, 7-10
Occupational Safety and Health Administration (OSHA), 186-87, 195
Office of Management and Budget (OMB), 129-31, 133, 134
Office of Naval Research (ONR), 125
"Old Pro," 32
On-site visits. *See* Site visits
Organizational structures, 26-28
Outside contractual services, 24-25

Patentability
abandonment of invention and, 108-9
novelty as requirement of, 107
of subject matter, 106-7
unobviousness as requirement of, 109-11
utility as requirement of, 107-8
Patent and Trademark Office (PTO), 110, 111
Patent attorneys, 106, 110

Patents
application process, 112-15
constitutional basis for, 104-5
determination of inventorship, 112
interference procedure regarding, 115-20
laboratory notebooks in application for, 120
legal issues, 110-11
ownership of, 105-6
priority establishment regarding, 116-17
trade secrets and, 102-3, 113
Paton, W. D. M., 228
Peer review system for research proposals, 157-60
Penicillin, 126
Performance variance, 43
Per se rule, 118-19
Personal observation as control technique, 54
Personnel. *See also* Equal employment opportunity and affirmative action
allocation of, 23, 24
employment guidelines, 212-15
as grant proposal element, 150-52
motivation of, 28-33
PERT (Program Evaluation and Review Technique), 47-50
PERT/COST, 50-51, 56, 57
PERT/CPM network, 12-14
Pesticides, 185
Pharmaceutical Manufacturers Association, 208
Plan document, 7
Planning. *See also* Short-range planning; Strategic planning
intermediate-range, 10-12
long-range, 4-5
Polaris Weapon System, 47
Pollution control, 192
Position design, 31-32
Posting documents, 162-64
Prior approval requirement, 164-65
PRISM (Program Reliability Information System for Management), 47
Private foundation grants, 141-43. *See also* Grant application process; Grants
Problems, 32-35
Professional publications, 88-90
Professional seminars, 90-92
Profile charts, 36-37
Program implementation
motivation of personnel, 28-33
organizational structure, 26-28

problem solving and decision making, 33-38
resource management, 21-26
Program project grants, 140-41
Progress reports, 166-67
Project abandonment, 55, 56
Project grants, 139
Projects
  budgets for, 15-17
  capital requirements, 17
  changes to, 17-18
  evaluation of, 11-12
  funding of, 10-12
  selection of, 10-11
Project teams, 26-27
Proposal Checklist and Evaluation Form, 144
Proposal planning. *See* Grant proposal process
Proprietary information, 101-2, 121. *See also* Patents; Trade secrets
Public Advisory Group, 157
Public Health Service (PHS), 165
  animal welfare policy, 232
Public Health Service Act, 184
Public relations, 96-98

Radiation Control for Health and Safety Act, 184
Radiation hoods, 74
rDNA. *See* Recombinant DNA
rDNA advisory committee, 196, 197
Reactivity, 189-90
Recission bill, 134
Recombinant DNA
  agency guidelines, 194-95
  current guidelines, 196-99, 202
  state and municipality guidelines, 196
Redier, R., 83
Reduction to practice, 116-20
Relevance trees, 55
Request for information (RFI), 175
Request for proposal (RFP), 175, 177, 178
Request for quotation (RFQ), 175-76
Research and development (R&D). *See also* Control techniques
  budgeting for, 15-17
  costs, 22, 23
  funding for, 10
  management of, 5
  mission of, 3
Research career development awards, 140, 141
Researchers, individual, 27-28

Research funds. *See also* Federal funding for research; Grants; Private foundation grants
  federal, 125-28
  governmental support, shifts in, 126-28
  management of, 21-23
  for projects, 10-12
The Research Grant Program of the Public Health Service, 157
Research organizations
  defining mission of, 3-4
  intermediate-range planning, 10-12
  long-range planning, 4-5
  short-range planning, 12-18
  strategic planning, 5-10
Research project grants, 140
Research reports, 89
Resource Conservation and Recovery Act (RCRA), 188
Resource grants, 139
Return on investment (ROI), 52, 57
Richardson, J. H., 75, 76, 83
Ripa, L. W., 218, 222
Rubber aprons, 72, 74
Rule-of-reason approach to patents, 119

Safety. *See* Laboratory safety
Safety audits and inspections, 69-70
Safety committee, 66-67
Safety equipment. *See* Equipment
Salaries, 23, 150-52
Scanlan, B. K., 21
Schmitz, T. M., 62
Science: The Endless Frontier, 126
Sears, Roebuck & Company, 101-2, 104
Sectional overhead, 22
Self-contained breathing apparatus (SCBA), 74
Seminars
  as employee incentive, 88
  speaking skills at, 90-92
Shannon, J. A., 127
Shapley, W. H., 128
Sharps, 80
Shields, 72, 74
Shoes for laboratory safety, 74
Short-range planning, 12
  budgeting in, 15-17
  capital requirements for, 17
  changes to plan in, 17-18
  computerized techniques, 15
  DELTA charts, 14
  Gantt techniques, 13-14
  networking, 12-13
Shulman, J. J., 94
Sick leave, 151

Site visits
  during contract selection evaluation, 180
  during grant proposal evaluation, 159-61
Slack paths, 47, 48
Slides, 35-millimeter, 93
Snyderman, B. B., 29
Social change, scientific knowledge and, 127
Social Security (FICA), 151, 152
Sole-source procurement, 175
Somerville, B., 156
Space resources, 26
Speaking skills, 90-92
Specialized objective control techniques, 44-50
Sperlinger, D., 227
Spill cleanup kits, 74
Spital, F. C., 55
Sputnik I., 126, 127
Starr, D. A., 162
Steere, N. V., 62, 67, 69, 73
Stiffel Lamp Company, 102
Strategic planning
  elements of, 5-7
  goals and objectives of, 7-10
Strategy managers, 6-7
"Study Section," 157
Subjective control techniques, 53-54
Supplies, 152
Support resources, 25-26
Supreme Court, U.S.
  on patentability, 106-7, 111
  on proprietary information, 101-2

Task groups, 27
Teague, G. V., 137, 179
Team members, 32
Technical communication, 87-88
  graphics and audiovisual materials used for, 92-94
  and obligations to public, 96-98
  in professional publications, 89-90
  for research reports, 89
  for seminar presentations, 90-92
  writing skills for, 94-95
Technical reports, 89
Technological forecasting techniques, 55
Technology transfer, 54-55
Theory X, 29, 31, 32
Theory Y, 29, 31
3,4-DCPA/3,4-DCAA controversy, 110-11, 114
Time-event analysis, 44
Time-plus-materials contract, 174

Time relationships, 13, 14
Time reporting, 22
Toxic chemicals, 78-79. See also Hazardous materials
Toxic Substances Control Act (TSCA), 185, 195, 230
Toxic wastes, 190, 191. See also Hazardous wastes
Trade secrets, 102-4
Transparencies, 93
Transportation
  of chemicals, 79-80
  of hazardous materials, 79-80, 189, 192-93
Transportation Department, U.S.
  guidelines on chemicals, 79-80
  guidelines on hazardous materials, 189, 192-93
Trash, 80
Travel expenses, 152-53
Treatment storage and disposal facilities (TSDF), 188-89, 191
Tregoe, B., 34
Trend extrapolation, 55
Trumble, R. R., 217
Tuskegee study, 217

Unemployment insurance, 151
Useful arts, 105
U.S. vs. Karl Brandt, 217

Vacation leave, 151
Variable costs, 52-53
Vasarhelyi, M. A., 50, 51
Virus, Serum, and Toxin Act, 195
Vocational Rehabilitation Act, 210

Wade, N., 183
Walters, J. E., 53
Waste chemicals, 81. See also Chemicals; Hazardous materials
Waste solvents, 81
Webber, D., 10
Weil, E. D., 5
White, V. P., 137
Wigodsky, H. S., 219
Williams, W. H., 227
Wood, W. S., 69
Workers' compensation, 151
Working alone, 83-84
World War II, 126
Wortman, L. A., 28

Xerographic photoreceptor case, 119
Xerography, 105